Augmented Vision and Reality

Volume 4

Series editors

Riad I. Hammoud, Kokomo, IN, USA
Lawrence B. Wolff, New York, NY, USA

For further volumes:
http://www.springer.com/series/8612

Fabiana Rodrigues Leta
Editor

Visual Computing

Scientific Visualization and Imaging Systems

Springer

Editor
Fabiana Rodrigues Leta
Mechanical Engineering Department
Universidade Federal Fluminense
Niterói, Rio de Janeiro
Brazil

ISSN 2190-5916 ISSN 2190-5924 (electronic)
ISBN 978-3-662-52029-1 ISBN 978-3-642-55131-4 (eBook)
DOI 10.1007/978-3-642-55131-4
Springer Heidelberg New York Dordrecht London

Softcover reprint of the hardcover 1st edition 2014

Printed on acid-free paper

Springer is part of Springer Science+Business Media (www.springer.com)

To my husband Anchizes
To my parents, Nicola and Ana Lucia,
in memoriam

Preface

Scientific Visualization and Imaging Systems include multidisciplinary areas, with applications in many knowledge fields such as Engineering, Medicine, Material Science, Physics, Geology, Geographic Information Systems, among others.

Scientific Visualization in this volume is understood as the science field that involves the visual representation of data acquired from experiments. As an expansion of this idea, it is also possible to include data examination generated by mathematical-physical modeling. Imaging Systems encompass any subject related to digital images, from fundamental requirements for a correct image acquisition to computational algorithms that make it possible to obtain relevant information for image analysis.

In this context, the International Conference on Advanced Computational Engineering and Experimenting (ACE-X) included in its program a Special Session on Scientific Visualization and Imaging Systems in 2010. This session aimed to stimulate discussion on researches that involve the use of digital images as an understandable approach for analyzing and visualizing phenomena, experiments, and cases. Since ACE-X 2010, which was held in Paris, paper submission considering any digital imaging topic and scientific visualization has been encouraged, which brings technological innovation on its methods and applications.

This book is a selection of 13 revised and extended research papers presented in the International Conference on Advanced Computational Engineering and Experimenting (ACE-X) Conferences 2010 (Paris), 2011 (Algarve), 2012 (Istanbul), and 2013 (Madrid). The examples were particularly chosen from materials research, medical applications, general concepts applied in simulations and image analysis, and other related interesting problems.

Although the book does not cover the entire universe of possible applications in Scientific Visualization and Imaging Systems, it presents important contributions that can benefit students and researchers that deal with visualization and imaging analysis.

Hence, I hope that all book chapters can be useful to those who are interested in developing research in such important field.

Fabiana Rodrigues Leta

Acknowledgments

I would like to thank all the authors for their participation and cooperation. The quality of their contributions was fundamental to the high-level outcome of this book.

This book became possible thanks to the confidence and support of Prof. Dr.-Ing. Andreas Öchsner, to whom I express my deepest gratitude.

I acknowledge the team of Springer-Verlag for the support in the course of the preparation of this book.

I thank all the friends who believed in this project, especially Mrs. Marcia Öchsner for her constant encouragement.

Finally, my deepest thanks go to my husband for his essential support, encouragement, and forbearance during all this important work.

Contents

Contributors

Rita Ambu University of Cagliari, Cagliari, Italy

Felipe de O. Baldner Instituto Nacional de Metrologia, Qualidade e Tecnologia—Inmetro, Duque de Caxias, RJ, Brazil

Diego C. Barboza Computer Science Institute, Universidade Federal Fluminense—UFF, Niterói, RJ, Brazil

Slobodan Marko Beroš Faculty of Electrical Engineering, Mechanical Engineering and Naval Architecture, University of Split, Split, Croatia

Mauro Biondi Mechanical Engineering Department, Computational and Dimensional Metrology Laboratory, Universidade Federal Fluminense—UFF, Niterói, RJ, Brazil

R. P. Bom Mechanical Engineering Department, Center for Science and Technology—CCT, University of the State of Santa Catarina—UDESC, Campus Universitário Avelino Marcante, Joinville, SC, Brazil

Sergio L. Braga Mechanical Engineering Department, Catholic University of Rio de Janeiro—PUC-RJ, Rio de Janeiro, RJ, Brazil

Esteban Clua Computer Science Institute, Universidade Federal Fluminense—UFF, Niterói, RJ, Brazil

Pedro B. Costa National Institute of Metrology, Quality and Technology, Duque de Caxias, RJ, Brazil

Cyril Drocourt Laboratoire des Technologies Innovantes, INSSET, Université de Picardie Jules Verne, ST Quentin, France

Frédéric Dubois Laboratoire De Mécanique et Génie Civil (LMGC), CNRS/Université Montpellier 2, Montpellier, France

Bruno Feijó Informatic Department, PUC-Rio, ICAD Games, Rio de Janeiro, RJ, Brazil

Jérome Fortin Laboratoire des Technologies Innovantes, INSSET, Université de Picardie Jules Verne, ST Quentin, France

João Gabriel Felipe M. Gazolla Computer Science Institute, Universidade Federal Fluminense—UFF, Niterói, RJ, Brazil

Dominique Gobin FAST—CNRS—Université Paris VI, Campus Universitaire, Orsay, France

Luca Goglio Politecnico di Torino, Torino, Italy

Juliana F. S. Gomes Instituto Nacional de Metrologia, Qualidade e Tecnologia—Inmetro, Duque de Caxias, RJ, Brazil

Jong-Eun Ha Department of Automotive Engineering, Seoul National University of Science and Technology, Seoul, Korea

Mark Joselli Computing Institute, Universidade Federal Fluminense-UFF, Niterói, RJ, Brazil

Dong-Joong Kang School of Mechanical Engineering, Pusan National University, Busan, Korea

Jeong-Hyun Kim Realhub Research Institute, Realhub Corporation Limited, Busan, Korea

Ivica Kuzmanić Faculty of Maritime Studies, University of Split, Split, Croatia

Fabiana R. Leta Mechanical Engineering Department, Universidade Federal Fluminense—UFF, Niterói, RJ, Brazil

Q. M. Li School of Mechanical, Aerospace and Civil Engineering, University of Manchester, Manchester, UK

P. Liatsis Department of Electrical and Electronic Engineering, City University London, London, UK

T. Lowe Henry Moseley X-ray Imaging Facility, School of Materials, University of Manchester, Manchester, UK

S. A. McDonald Henry Moseley X-ray Imaging Facility, School of Materials, University of Manchester, Manchester, UK

Giangiacomo Minak University of Bologna, Bologna, Italy

E. H. Murakami Whirlpool S.A. Compressor Plant, Joinville, SC, Brazil

Marco Peroni EC Joint Research Centre, IPSC Institute, Ispra, VA, Italy

Maria S do Souza Mechanical Engineering Department, Computational and Dimensional Metrology Laboratory, Universidade Federal Fluminense—UFF, Niterói, RJ, Brazil

Y. L. Sun School of Mechanical, Aerospace and Civil Engineering, University of Manchester, Manchester, UK

Zhu Teng School of Computer and Information Technology, Beijing Jiaotong University, Beijing, China

Harold Trannois Laboratoire des Technologies Innovantes, INSSET, Université de Picardie Jules Verne, ST Quentin, France

Luis Valente Computing Institute, Universidade Federal Fluminense-UFF, Niterói, RJ, Brazil

Jakson Manfredini Vassoler Universidade Federal do Rio Grande do Sul, Porto Alegre, Rio Grande do Sul, Brazil

Gisele Maria R. Vieira Mechanical Engineering Department, Federal Center of Technological Education Celso Suckow da Fonseca—CEFET/RJ, Rio de Janeiro, RJ, Brazil

Dean Vucinic Mechanical Engineering Department, Department of Electronics and Informatics, Vrije Universiteit Brussel, Brussels, Belgium

Mirjana Vujović Occupational Medicine Private Practice, Ploče, Croatia

Igor Vujović Faculty of Maritime Studies, University of Split, Split, Croatia

Y. Wang College of Astronautics, Nanjing University of Aeronautics and Astronautics, Nanjing, China

P. J. Withers Henry Moseley X-ray Imaging Facility, School of Materials, University of Manchester, Manchester, UK

Marcelo Zamith Computing Institute, Universidade Federal Fluminense-UFF, Niterói, RJ, Brazil

Andrea Zucchelli University of Bologna, Bologna, Italy

Part I
General Concepts Applied in Simulations and Image Analysis

Chapter 1
Multidisciplinary Scientific Visualization in European R&D Projects

Dean Vucinic

Abstract The manuscript addresses Scientific Visualization (SV) challenges in the Modeling and Simulation (M&S) environments, experienced by the author in the multidisciplinary European Research and Development (R&D) projects, which are important to be resolved for the growing complexity in evolving engineering software and their related visualization features, as an essential factor to improve their quality and efficient use. The visualization tools are considered as the natural integration mechanism to evidence the complexity and to provide missing integration solutions. Today, the visualization tools are equipped with highly interactive visual aids, which allow analysis and inspection of complex numerical data generated from high-bandwidth data sources such as simulation software, experimental rigs, satellites, scanners, etc. Such tools help scientist and engineers in data extraction, visualization, interpretation, and analysis tasks, enabling them to experience high degree of interaction and effectiveness in solving engineering problems. The modern engineering design is, and has to be based on the M&S principles, the key enabler to combine multidisciplinary workflows, which manage and structure such highly complex industrial solutions, supposed to integrate generic M&S tools based on the open standards solutions. The author gained experience in solving the M&S problems in European engineering R&D projects is presented throughout a time-span of last two decades, where the software technologies have been researched and demonstrated, especially addressing the SV software evolution in engineering, and in addition, the information visualization in general.

Keywords Scientific visualization · Modeling and simulation · Multidisciplinary workflows · Information visualization · European research and development projects

---

D. Vucinic (✉)
Department of Mechanical Engineering, Department of Electronics and Informatics, Vrije Universiteit Brussel, Pleinlaan 2, 1050 Brussels, Belgium
e-mail: dean.vucinic@vub.ac.be

F. R. Leta (ed.), *Visual Computing*, Augmented Vision and Reality 4,
DOI: 10.1007/978-3-642-55131-4_1,

List of Abbreviations

CAD	Computer aided design
CAE	Computer aided engineering
CFD	Computational fluid dynamics
CFView	Computational flow field visualization
EFD	Experimental fluid dynamics
EU	European union
GUI	Graphical user interface
IME	Integrated modeling environment
J2EE	Java 2 platform, enterprise edition
JOnAS	Java open-source J2EE application server
KBE	Knowledge based engineering
KDD	Knowledge discovery and data mining
MVC	Model View Controller
MVE	Modular visualization environments
OOM	Object oriented methodology
OOP	Object oriented programming
OOPL	Object-oriented programming language
PIV	Particle image velocimetry
QFView	Quantitative flow field visualization
SDK	Software development kit
SOAP	Simple object access protocol
SV	Scientific visualization
SW	Scientific workflow
VisAD	Visualization for algorithm development
VA	Visual analytics
VTK	Visualization toolkit
VUB	Vrije Universiteit Brussel
WWW	World Wide Web
WS	Web services

1.1 Introduction

The Scientific Visualization (SV) software [1] is empowering the knowledge discovery [2, 3] in the continuously evolving simulation-based engineering practice [4–6]. The chapter addresses the multidisciplinary visualization aspects from the computer science perspective of SV, envisaging that the novel generation of engineering software environments will provide more intelligent visualization components, with the possibility to access and to reuse the required engineering multidisciplinary knowledge and technologies in two different situations to aid:

1. software development process of such engineering software tools and later on
2. engineering best practices, as integrated part of such software tools, enhancing the end-user functionality.

1.1.1 Modeling and Simulation

The SV software has established itself as an integral part of the Modeling and Simulation (M&S) software tools [7, 8].

The model or better saying multiple of them—the models, are the core data elements in the multidisciplinary M&S engineering software, enabling the simulation—a process of approximating the real-world things (products, events…), by covering the three fundamental elements in the M&S process [9]:

1. Modeling—creating the models
2. Simulation—approximating models behavior
3. Analysis—drawing conclusions through models verification and validation

In which the above-mentioned basic elements are tightly coupled with the SV components in order to sustain:

- ability to represent models,
- way to interface with models,
- M&S as the problem-based discipline that allows for the repeated testing of the investigated hypothesis.

SV presence, providing the advanced visualization functionality through the introduced M&S precepts, is expected to underpin the future research and new products developments, when treating multidisciplinary engineering applications domains.

1.1.2 Collaborative Environments

The collaborative environments [10] are becoming indispensable working instruments, which will draw from and integrate together a large variety of different scientific disciplines, professions, researchers, and engineering technologies.

This will be achieved by bringing together their methods and insights, in an easy-to-use way and in an operationally efficient manner. In the example shown in Fig. 1.1, different expert groups present in the aircraft design process involve together a variety of different engineering domains. They need interactions and exchange of pertinent information in order to come up with an appropriate aircraft design, which has to fulfill the variety of multidisciplinary requirements. The 3D visualization models are often used across such multidisciplinary teams to demonstrate and visualize the envisaged solutions in a focused problem-oriented way. An important requirement for such collaborative visualization environment is the

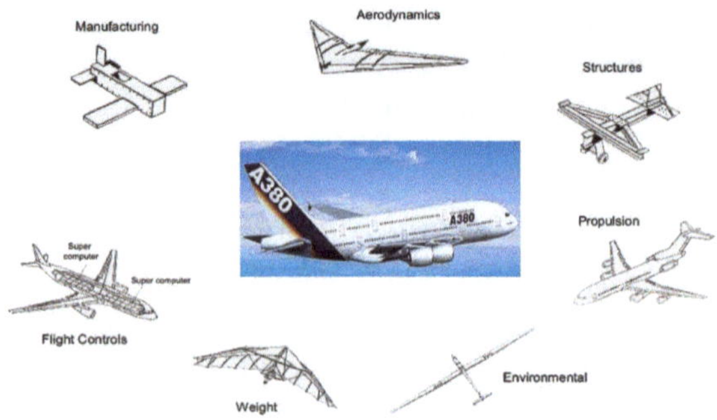

Fig. 1.1 Multidisciplinary domains present in the aircraft design process [11]

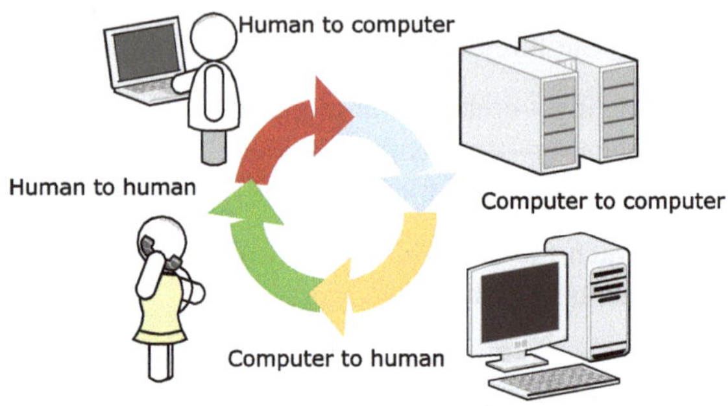

Fig. 1.2 Human–computer communication types

interoperability of data sources created and manipulated within such utilization design scenario. It is essential that the data exchange and sharing of information underpin the decision-making process through the human–computer interactions, see Fig. 1.2, in the following ways [12]:

(1) Human-to-human: teaching, research, thinking and defining what to consider.
(2) Human-to-computer: specifying the problem in order to find appropriate service/application.
(3) Computer-to-computer: service/application discovery, utilization, and monitoring.
(4) Computer-to-human: feedback through graphical/textual display of processed information.

The problem of combining/integrating different software tools to solve multidisciplinary problems is faced with the increase in complexity of the involved data models (Fig. 1.1).

Thus, we need appropriate software development tools and applications, which can enable the reuse and impose the correct usage of the inherent knowledge, contained, not only within the software tools, but also used by the experts employing them, when proposing new solutions. As described in [13], we have to advance our research in such direction, by introducing knowledge (ontology) concepts for the new software solutions, providing intelligent execution and treatment of the involved

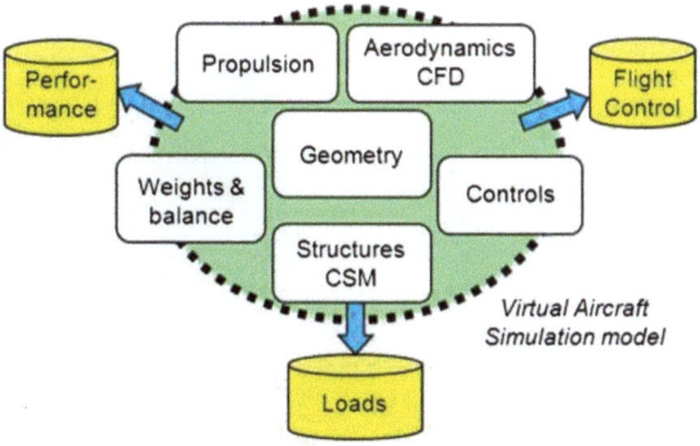

Fig. 1.3 Virtual aircraft M&S software [14]

visualization and presentation tasks. These tasks are crucial for gaining the thrust and confidence in the accomplished results across multidisciplinary domains, thus as "seeing" visualize them, makes the important missing difference. For example, the European Software tool CEASIOM [14] is meant to support engineers in the conceptual design process of the aircraft, with emphasis on the improved prediction of stability and control properties achieved by higher fidelity methods found in the contemporary aircraft design tools (Fig. 1.3).

1.1.3 Specialized Engineering Tools

Within the virtual aircraft design environment, a Computational Fluid Dynamics (CFD) engineering tools are found to support the calculation of the aircraft aerodynamics behavior. An example, with more details, is shown in Fig. 1.4, for the wing aerodynamics simulation.

In the mentioned CFD example, the engineering tasks may lead to locate a flow recirculation area, a region of high turbulence, or the points where the value of enthalpy reaches its minimum and maximum values. Usually, the goal is not only to see the flow pattern, but also to understand why and how it develops. The possibility to apply several visualization tools in an interactive way makes it possible to uncover the problem under investigation. The exploration of the numerically generated data leads to the concept of virtual instrument, namely a software probe that can be interactively manipulated.

SV methods empower engineers/researchers to analyze the same problem in a variety of ways. They promote the concept of virtual laboratory where the user performs virtual measurements. These virtual measurements are the results of the interactive visualization process that the user applies on the analyzed models.

An example is the case for CFView [15], see Fig. 1.5, an SV application developed by the author over the 1988–1998 period, which is today a part of the "FINE", NUMECA's engineering environment. Some of the known visualization applications for CFD and Finite Elements Analysis (FEA) are EnSight from CEI

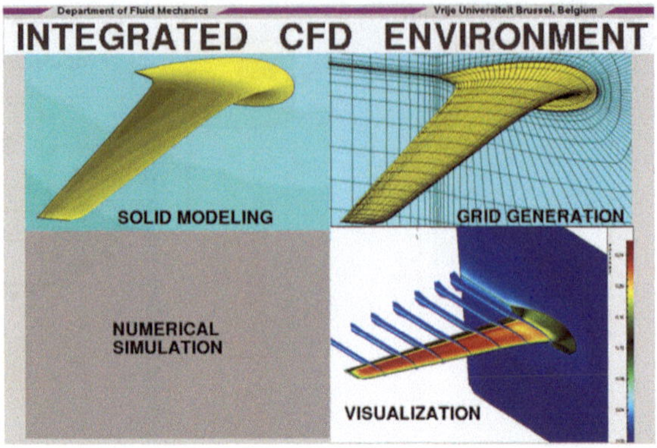

Fig. 1.4 Integrated CFD software

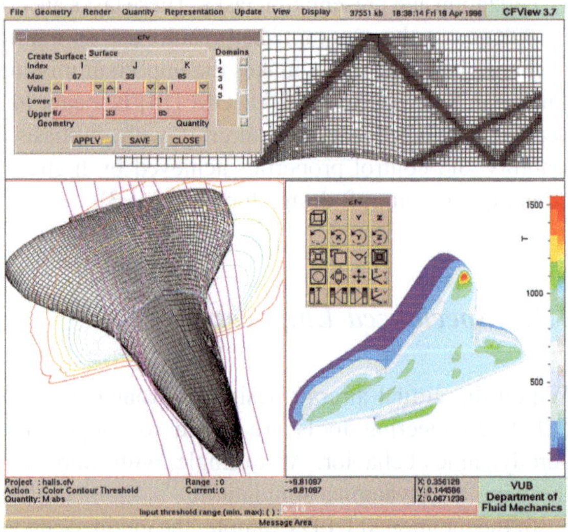

Fig. 1.5 CFView the scientific visualization system

[16], FieldView from Intelligent Light's [17, 18], TecPlot from Amtec Engineering Inc. [19], ParaView [20], Mayavi [21], and VisIt from Lawrence Livermore National Laboratory [22]. Such programs are appropriate for users who need a specific SV functionality, and when considering a multidisciplinary environment the emerging problem is how such tools can be efficiently integrated in a multiapplication engineering environment in order to that each of their internal domain knowledge can be automatically reused and shared by other applications and human actors.

1.1.4 Visual Analytics

The mentioned CFD analysis process stresses the importance of only one of the aspects found in the previously illustrated aircraft design process. Such consideration brings directly forward the CFD specific domain knowledge, in which the

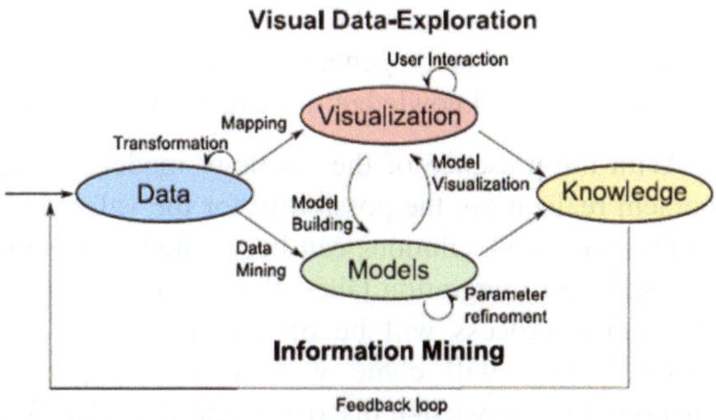

Fig. 1.6 Visual analytics for knowledge discovery

appropriate SV representations are required to be automatically involved, when such task is executed as part of the multidisciplinary investigation problem. The goal is to find an appropriate way to integrate the relevant domain knowledge, in order to become part of the envisaged integrated multidisciplinary knowledge base. It seems that an appropriate software solution can be found in Visual Analytics (VA) methodology [23, 24], which combines SV with the information data mining, emerging as an important research line to follow, see Fig. 1.6. VA is enlarging SV, and can be considered as an extended approach in the developments of multidisciplinary visualization features by integrating data analysis know-how. Especially, the VA research in the human–computer interaction is focusing on: collaboration, interaction, presentation, and dissemination, based on utilizing knowledge representation for knowledge discovery.

The software development process of such complex multidisciplinary software requires the knowledge engineering approach, and more specifically the ontology modeling [25], where each software component is expected to have its own ontology model and the respective knowledge base associated to it. For example, there will be the possibility to query the data model structure and its related functionality found in the applied software components, but in a machine programmable way, thus avoiding end-user input. The expected benefit of such approach is that the software integration will become more automatic and less error prone, when achieved by interrogating ontology and the related knowledge by the program-to-program, or machine-to-machine based way. Such integration process will take control over the data and processes, through the well-defined functionality, which will be enriched with the relevant semantics. These semantics components will enable the applications and users to explore/reuse/enlarge the domain engineering knowledge—present within applications (software-coded) and will continuously evolve in such future software solutions.

The basic design principle for the ontology modeling is to support the features that are necessary for a specific application domain (user-engineer point of view). The ontology models represent the domain concepts and their relationships, thus they define the domain language (semantics) meaningful to humans and usable by machines, in order to improve:

1. (semantic) search and browsing
2. integration of heterogeneous information sources
3. analytics and knowledge discovery capabilities

With the presence of the ontology models in such envisaged software development realization, the possibility for the validation and the compatibility testing of the integrated solutions can be verified in an upfront manner; even before the real software integration takes place. It is expected that such envisaged software integration process will be of a better quality, as the involved concepts and methodologies will come with a more explicit and declarative knowledge, enabling the knowledge discovery and data (KDD) mining.

1.1.5 Ontology and 3D Graphics

The problem in associating the semantic information to 3D models can be solved by applying the two web-based software standards: X3D (eXtensible 3D—X3D Graphics) [26, 27] and Semantic Web [28] (Fig. 1.7).

The main idea of the Semantic Web is to delegate many current human-specific activities to computers, and therefore data need to be presented in a machine-readable format. Since XML (eXtensible Markup Language) is intended for describing the meaning of Web data (or, in other words, their semantics), most of the current Semantic Web technologies belong to the XML family and the respective layers combining the Uniform Resource Identifier (URI), Namespaces, and Unicode technologies are also XML based. The XML models are the essential part of the Semantic Web as they do not embrace only the domain-specific content (such as design data) but also the "technological" aspects modeled also as XML (such as Extensible Stylesheet Language Transformations (XSLT) [29], Resource Description Framework (RDF) [30], or Web Ontology Language (OWL) [31]) (Fig. 1.8).

Ontologies represent the fundamental models for the Semantic Web implementation, as they offer appropriate mechanisms to describe relationships between objects and properties, as for example not found in X3D.

The multidisciplinary information visualization is inherently complex, and to enable the efficient interaction with 3D/4D space–time domain, the graphics content of the created and displayed models could be enriched with a domain-specific knowledge, in order that users can effectively query, interpret, personalize, and manipulate the visualized information. The semantic enrichment of 3D models plays important roles as follows: (a) enables users to understand and interact with the complex and sometimes incomprehensive displayed representations, (b) underpin the creation of the user-friendly and ergonomic Graphical User Interface (GUI), and (c) providing a knowledge base for the intelligent queries.

To demonstrate the mentioned approach, an example of the multiphysics fluid–structure interaction (FSI) problem was chosen, as combines the numerical

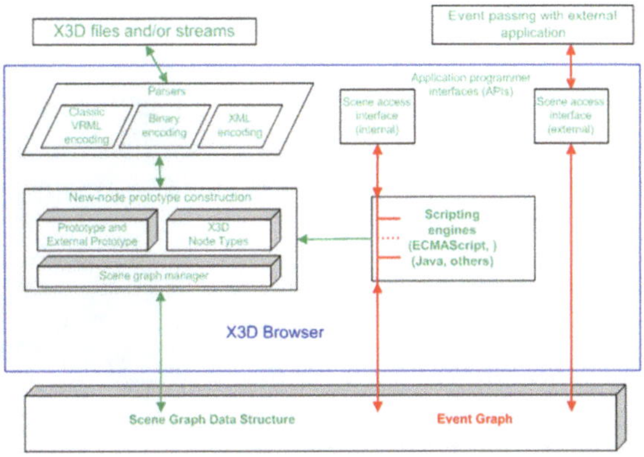

Fig. 1.7 X3D architecture [26]

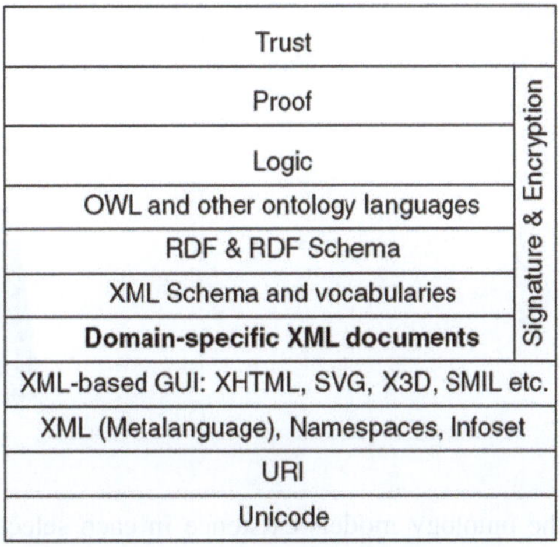

Fig. 1.8 Architecture of the semantic Web

modeling of the fluid and solid continuum, respectively CFD (Fig. 1.9) and FEM (Fig. 1.10) simulation results are shown. In order to enable integration and interoperability of such software tools, we have to define ontologies (metadata descriptors), which explicitly define numerical models, concepts, and their definitions.

As our ontology-based approach is evolving from the object-oriented modeling [32], we understand ontologies as an explicit knowledge about the domain concepts, in our case multidimensional numerical models, designed and implemented as a set of classes (objects) used in software development. An example of detailed object-oriented approach for developing SV system is given in [33].

The FSI problem is complex, and thus the use of ontologies is expected to formally reveal the applied engineering models in order to that the interfacing process between FEM and CFD software can be automatically achieved through software-to-software communication. We expect that the suggested ontology approach will improve interoperability and integration of such tools by requiring

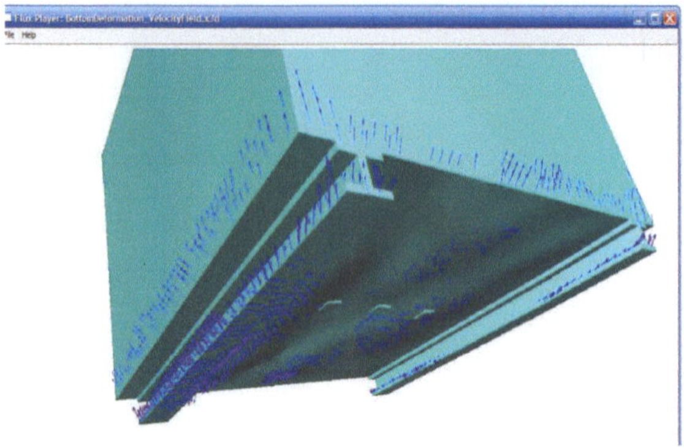

Fig. 1.9 CFD model in X3D

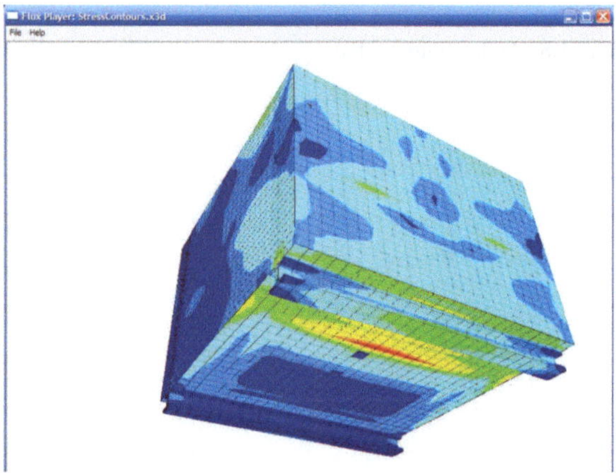

Fig. 1.10 FEM model in X3D

the ontology models existence in each selected software tool, and thus prior to their integration. Such ontology-modeled knowledge is expected to make visible the intrinsic parts of the involved software applications enabling the application-to-application communication. The expected benefits are in the reduction in the engineer's (domain experts) time when using the software and in the software development efficiency by eliminating programmable errors, introduced when programmers have to program manually such complex software integration.

The envisaged approach is expected to contribute:

- Coherent semantics: different application domains develop different terminology for describing the same concept, but for example, when the results are graphically displayed, it is far easier to agree on them.
- Knowledge reuse: there is a need for the development of related ontologies outside the application tools and data models, in order to be able to reuse the exiting know-how in-between multidisciplinary software components.

Conceptually, each X3D application is a 3D time-based space of graphical, video, and audio objects that can be loaded, exchanged, and reused in a distributed

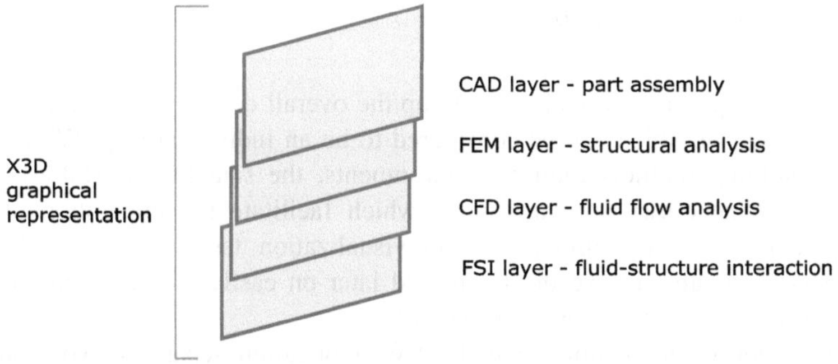

Fig. 1.11 Integration of 3D graphics with X3D layers

computer environments. In addition, the X3D objects can be dynamically modified through a variety of available mechanisms [26]. The semantics of X3D describes an abstract functional behavior of time-based interactive 3D and multimedia information. As shown in Fig. 1.7, each X3D application can:

- establish implicitly a world coordinate space for defined objects, as well as the objects included by the application;
- define and compose explicitly a set of 2D, 3D, and multimedia objects;
- specify hyperlinks to other files and applications;
- model objects behaviors;
- connect to external modules or applications via programming and scripting languages.

With combining the domain-specific and X3D-based ontologies we can create different 3D models for engineering purposes, see Fig. 1.11. The engineering models are interrelated using the ontologies shared semantics, which enables the developments of a flexible environment infrastructure to facilitate the semantics adaptation to multidisciplinary scenarios, as presented by CFD and FEM layers.

The semantic "model" is a concise representation of concepts that allow investigations of its properties, and in addition it enables an unambiguous representation of different domains viewpoints. A model has to clearly convey what it is about and defining the concept being depicted, in order that the stakeholders can unambiguously identify their viewpoints. However, for the model to be applicable, modelers should avoid to create fully comprehensive models defining a complete set of attributes. The value of a model resides in its ability to focus on the concern at hand, and to reduce complexity by eliminating the nontangible elements. The hiding of pointless and redundant data is the key of making an efficient ontology. Thus, we find Semantic Web an appropriate vehicle to obtain the knowledge about objects, by presenting and using only the parts of that available knowledge, which suits our demand.

1.1.6 Scientific Workflows

The engineering teams typically divide-up the overall design process into smaller subtasks, each of which can be considered to be an individual step. Within such problem-solving multidisciplinary environments, the scientific workflows (SW) [34] are the necessary software tools, which facilitate the integration of data management, analysis, simulation, and visualization task, as they enable the engineers to create and execute them, and later on easily modify such complex engineering tasks in an automated manner.

It is important to mention that the SW application separates two modeling concerns: component communication and overall workflow coordination. It has been proposed [35] that the separation of the conventional data modeling (structural data type) and the conceptual modeling (semantic type) gives rise to an advance SW design, where the validation of SW, as well as the discovery of type-conforming workflow implementations via replacement rules and by inserting appropriate semantic and structural adapters for workflow integration, prove that the related semantics presence through ontology modeling has become an indispensable part in the software development of such multidisciplinary tools.

For ontology creation, the Protégé [21], an open-source platform, is applied. It comes with a suite of tools for the construction of domain models and knowledge-based applications using ontologies (Fig. 1.12). Protégé can be customized, to provide domain-friendly support for creating knowledge models and their population. Protégé has extendable plug-in architecture and a Java-based Application Programming Interface (API) for building knowledge bases and related applications.

The multiple layers present in the described FSI workflow, usually are not needed to be modeled all. If the goal is the CAD part assembly model, which takes into account the stress deformations coming from FEM, the workflow will consist of the CAD layer and the FEM layer. If the goal is the product production, which needs only the CAD assembly parts, the workflow will be reduced to the CAD layer. If the goal is to solve the FSI problem, the FEM and CFD layers need to be involved. In order to describe such SW, the necessary ontology models need to be combined, requiring that each layer comes with its defined ontology [36] (Fig. 1.12). In our example, the four ontologies are mapped to the X3D ontology, which is the common ontology and thus the integration of these layers can be semantically validated, resulting that such SW can be enabled. In addition, each layer present as a separate X3D model together with the overall scene bringing all them together may represent the final end-user result, which also results in the validated software integration process, when creating such SW (Fig. 1.13).

SW [37, 38] has emerged as a useful paradigm to describe, manage, and share complex scientific analysis data (Fig. 1.14). SW represents declaratively the components or applications that need to be executed in the complex M&S software environment, as well as model dependencies among those components.

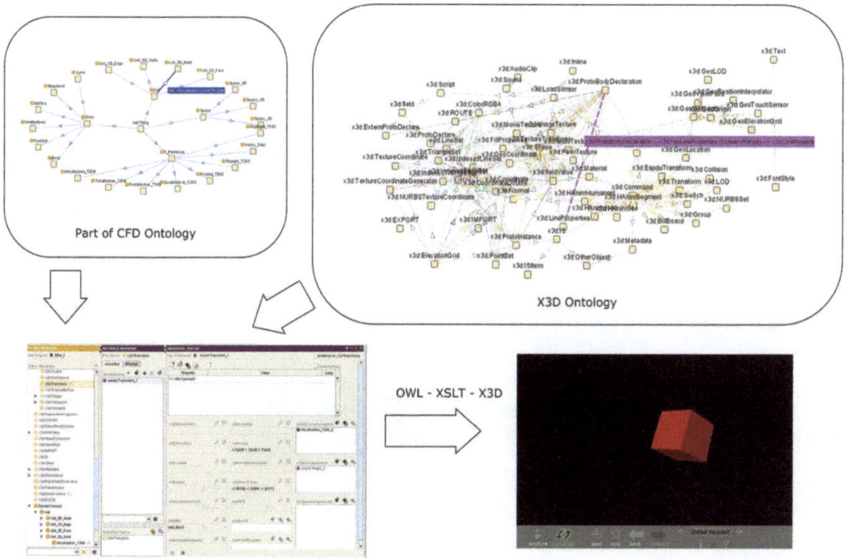

Fig. 1.12 Modeling CFD ontology with Protégé

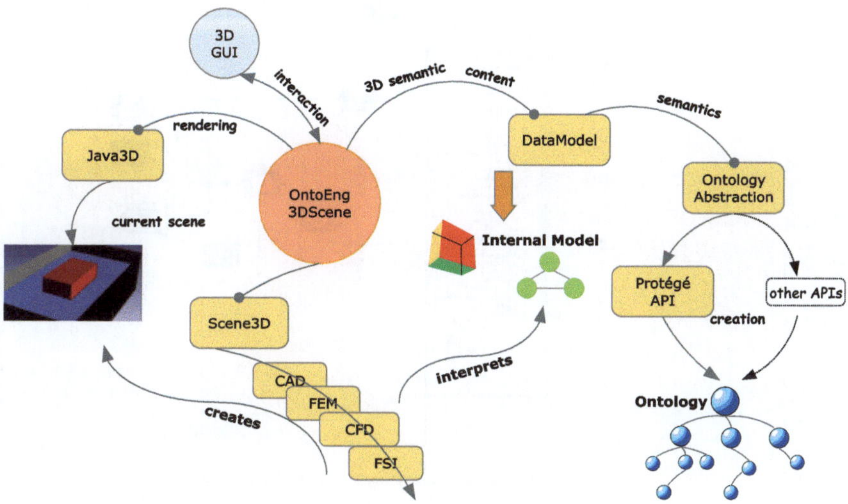

Fig. 1.13 Towards ontology-based software architecture

1.1.7 Introduction Summary

As discussed, the mentioned technologies can provide appropriate solutions to support the advancement of the SV aspects in improving M&S software by solving the following issues indicated as examples:

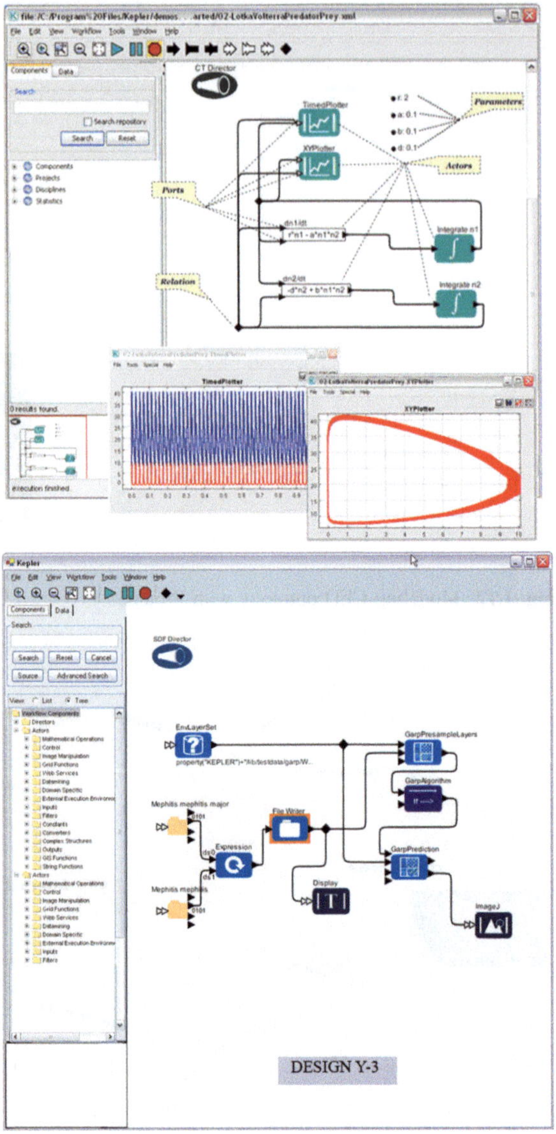

Fig. 1.14 Kepler scientific workflow software [37]

1. How to build/select data models that can be effectively used by the visualization components to present the information correctly to the users, while offering to them tools for real-time interaction in a natural and intuitive manner.
2. How to include high-resolution wall-displays connected to high-performance networked computing resources.

Such systems and architectures are no longer a mere vision: they are becoming reality, which opens new challenges for SV software researchers and developers, see Fig. 1.13. These are very well understood as important developments and thus such research topics are continuously being funded within the European R&D projects, which the author has been able to experience during the last two decades. In the following chapter some of these projects are addressed, highlighting their multidisciplinary visualization challenges.

1.2 European R&D Projects

In 1988, the author started the development of the SV system that was named "CFView" [39], and the author has established the Object-Oriented approach [32] for developing SV systems. The object-oriented programming language C++ was chosen, because of its high performance in number crunching and its capability to support OOP constructs.

CFView [15] was designed to work with structured and unstructured mesh types. CFView was found to be particularly valuable when treating the CFD simulations of the European Hypersonic Database [40] and validating the simulations against experimental data. CFView allows the simultaneous inspection of fluid flow simulations with structured and unstructured grids; as it uses data extraction algorithms such as plane-cutting, iso-surface, and particle-tracing for both grid models, to uncover interesting features hidden in such large datasets. These routines require large computational resources and tend to limit the interactive feedback of CFView. In order to overcome this problem, a heterogeneous system named "Parallel CFView" was constructed which distributes the computing load over several processors and permits real-time interactive visualization of CFD data. The development of Parallel CFView [41] was part of the EU-funded projects PAGEIN [42] and EEC/ESPRIT PASHA [43].

In the PASHA project, CFView visualization system was ported to MIMD and SIMD platforms for comparative benchmarking. The performance of the MIMD implementation was evaluated on two industrial test cases provided by the EUROPORT-1 project. Both SIMD and MIMD parallel machines were used to provide massive back-end processing power. A key achievement of the Parallel CFView project turned out to be the interface for communication between machines with heterogeneous architectures. An analysis was carried out to compare the performance of such distributed computing environment consisting of both: the SIMD and MIMD implementations and the sequential CFView system. The results showed that the parallel implementation of the extraction algorithms speedup the performance of the visualization process by a factor of 10, which proved the viability of such visualization system configurations.

PAGEIN [42] demonstrated more effective resource utilization through the support of distributed computing and collaborative engineering. The goal was to seamlessly integrate scarce human expertise in domains like, supercomputers, massively parallel computers, and data servers. PAGEIN enabled the engineers to exploit the computing resources and reach the critical expertise mass without need for colocation. For the users PAGEIN has integrated on top of wide-area networks, applications, and tools for supercomputing, visualization, and multimedia. It has demonstrated the viability of such combined technologies in CFD to be exploited on a European scale by aerospace actors.

The Live Code Learning Multimedia System (LCLMS) [44] was the IWT founded project, in which the database and network aspects for multimedia datasets were researched. This project strongly influenced the author's own research toward distributed and collaborative environments for engineering applications based on low-cost PC-platforms. LCLMS provided the basic technology upon which QFView was designed.

1.2.1 Alice: QFView Toward the Transparent Visualization of Numerical and Experimental Data Sets

The development of the Quantitative Flow Field Visualization (QFView) system [45] was made in the ESPRIT Project 28168 ALICE. QFView is a distributed software environment that integrates EFD and CFD data processing, including flow field mappings with flow field visualization, see Fig. 1.15. QFView was devised to support a unified treatment of data while providing for:

- the validation of results from experimental and numerical systems,
- the archival and retrieval of data from unified (experimental and numerical) flow field database.

Based on proven Internet and World Wide Web (WWW) standard technologies, QFView provides an integrated information system for fluid dynamics researchers. QFView is a web-based archival, analysis and visualization system, which enables the manipulation and extraction of data resulting from laboratory measurements or computational simulations. The system is suited for combining experimental and computational activities in a single operational context. This results in an increase of productivity, since the system facilitates for the exchange of information between investigators who conduct the same or similar simulations/experiments in different geographical locations, can be conducted in collaboration or independently.

The development of QFView in the ESPRIT-IV "ALICE" project (EP-28168) extended the author's research toward using the World Wide Web for designing and building up distributed, collaborative scientific environments [45, 46]. QFView was developed in a web-oriented client–server architecture (e.g., Java, JDBC) which allowed openness and modularity, as well as improved flexibility and integration of the visualization components (current and future). A core element was the creation of a central database where very large datasets were imported, classified, and stored for reuse. The distributed nature of QFView allows the user to extract, visualize, and compare data from the central database using World Wide Web access. QFView integrates EFD and CFD data processing (e.g., flow field mappings with flow field visualization).

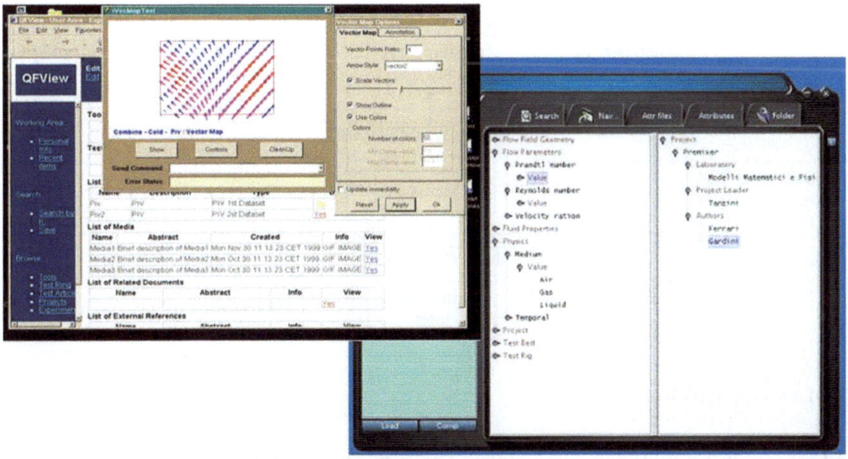

Fig. 1.15 QFView web Interface

QFView supports the integrated use of visualization and animation tools which are integrated with the database management system for data access and archiving. For example, in a possible configuration of QFView components over the Internet, the J2EE application servers [47], and the Database Management System servers (DBMS) [48] are located in one site, while the GUI client applications run on a server in another location (Fig. 1.17a). The meta database is based on a J2EE distributed architecture whose execution logic is stored and executed at the EJB container level [49], and which is accessed from the Java-based graphical user interface (GUI) via the http protocol. The main advantage of the Internet is the possibility to store and access data (images, input files, documents, etc.) from any URL in the world. Both EFD and CFD applications generate huge quantities of data which can be used to populate the database. The user needs to access, extract, visualize, and compare the required quantities; these functions as illustrated in Fig. 1.16. The QFView system is composed of three major elements:

1. An "EJB container" with all the metadata management rules to manipulate metadata and the relational database used to store the metadata information. The EJB container acts as security proxy for the data in the relational database.
2. A "Thin GUI Java client" is used for remote data entry, data organization, and plug-in visualization. GUI clients must be installed at the end-user location, either at application installation time or by automatic download (Zero Administration Client).
3. URL accessed data (images, video, data files, etc.) can be placed at any URL site.

QFView organizes stores, retrieves, and classifies the data generated by experiments and simulations with an easy-to-use GUI. The data management component (Fig. 1.17b), offers a user-friendly web-enabled front-end to populate

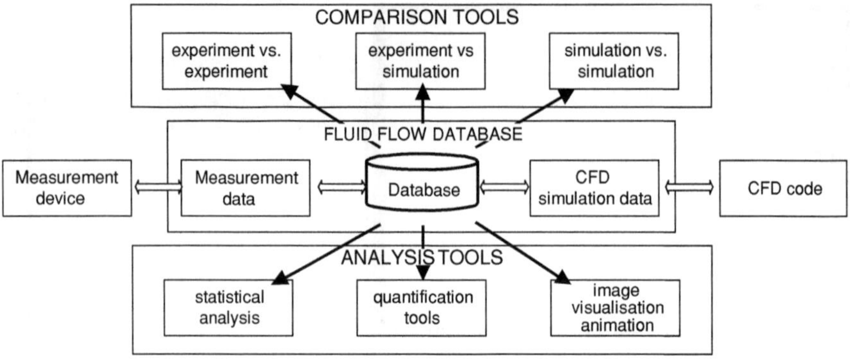

Fig. 1.16 The QFView framework

and maintain the metadata repository. The user can accomplish the following tasks using this thin GUI Java client:

- Start the GUI Java client application, create a new folder (measurement), define metadata information for the folder, such as keywords, physical characteristics, etc. It is important to emphasize that the GUI Java client is connected to the EJB server using the http protocol, and all the information entered by the user are automatically stored in the relational database.
- Organize the data into a hierarchy of predefined or newly created tree-like nodes; user can also execute a data search procedure, combine documents, and do several other operations on the input folder.
- Create and define new raw data such as XML files, images, input files, etc. for a particular folder by specifying either a local or a remote location of the data.
- Define old raw data (XML files, text files, videos, etc.) by specifying either a local or a remote location of the data.
- Start the visualization plug-in application on the selected raw data.

Essential Database Tools present in QFView are:

(a) Data entry
(b) Folder and document insertion
(c) Data classification
(d) Full text search
(e) Meta model and data organization

These tools were used to assist experimental setup, to improve data acquisition and systematic manipulation of the results of the VUB's double-annular jet modeling (Fig. 1.18, frame 1 from left) of the flow in a prototype combustion chamber. The requirement was to analyze the flow from LSV data (frame 2), PIV data (frame 3), LDV data (frame 4), and CFD calculation data (frame 5). The LDV (frame 4) clearly shows the mean flow field, which can be compared with CFD (frame 5). Qualitatively they produce a similar topology but the prediction is not

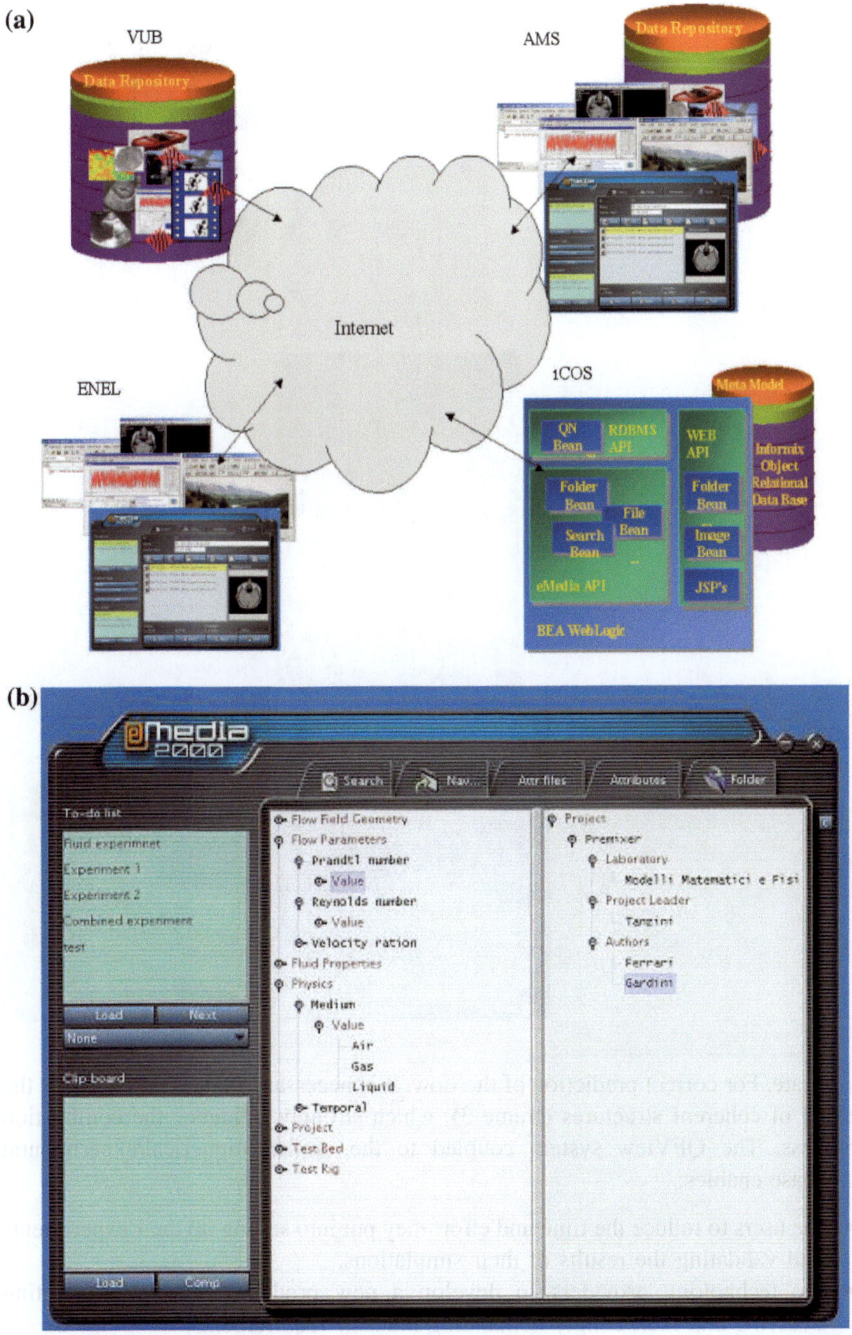

Fig. 1.17 QFView—an internet based archiving and visualization environment. **a** Distributed environment. **b** Web-based data management component

Fig. 1.18 VUB Burner experiment

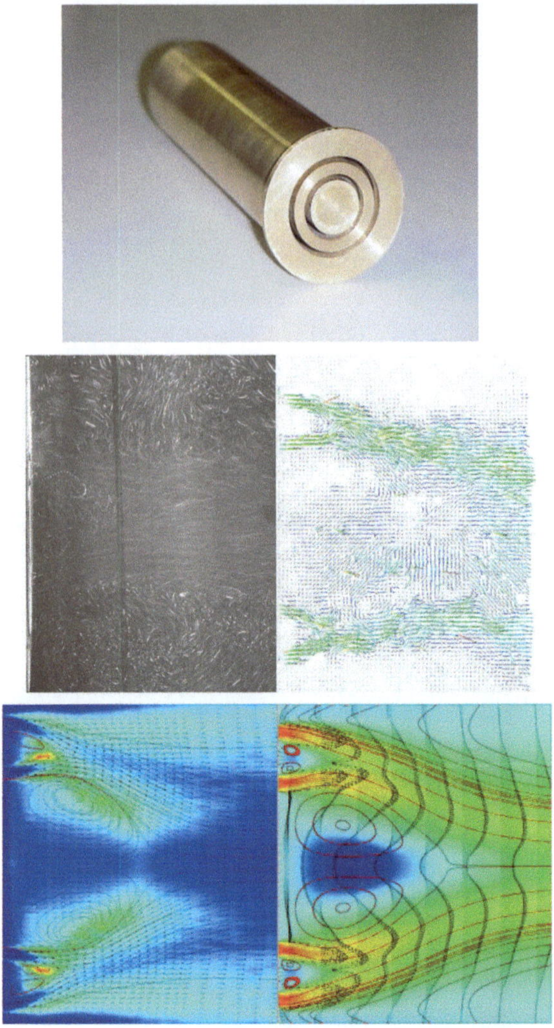

accurate. For correct prediction of the flow, it is necessary to take into account the effect of coherent structures (frame 3), which strongly influence the combustion process. The QFView system coupled to the unified numerical/experimental database enables:

- the users to reduce the time and effort they put into setting up their experiments and validating the results of their simulations,
- the technology providers to develop a new products capable of meeting evolving and increasingly demanding industry requirements.

The users have observed that QFView provided them with a means not only for archiving and manipulating datasets, but also for organizing their entire workflow.

The impact of visualization systems like QFView on the investigative process itself opens the way to entirely novel ways of working for researchers in experimental and computational fluid dynamics. Integrated, distributed, collaborative and visualization environments give the possibility of and indeed point to the need of reorganizing research methods and workflows; the various "experiments" conducted with QFView in ALICE have given glimpses of the exciting advances that one may expect from such systems in the coming years.

1.2.2 QNET-CFD: Multidisciplinary Knowledge Base

QNET-CFD was a Thematic Network on Quality and Trust for the industrial applications of CFD [50]. It provided European industries with a knowledge base of high-quality application challenges (reviewed and approved CFD results) and best practice guidelines. QNET-CFD was part of the EC R&D GROWTH program.

The main objectives of the project were to create the knowledge base to collect Application Challenges (AC) and Underlying Flow Regimes (UFR) from the trusted sources and made them available to the European industries, see an example Fig. 1.19.

The knowledge base is structured around the following six Thematic Areas (TA) and four Underlying Flow Categories:

Thematic Areas

TA 1 External aerodynamics
TA 2 Combustion and heat transfer
TA 3 Chemical and process, thermal hydraulics and nuclear safety
TA 4 Civil construction and HVAC
TA 5 Environmental flows
TA 6 Turbomachinery internal flows

Underlying Flow Categories

1 Free flows
2 Flows around bodies
3 Semi-confined flows
4 Confined flows

As illustrated in the Fig. 1.20, all the application challenges are associated with a single TA or industry sector. The UFRs are linked to ACs across the thematic areas. This is an important feature of the QNET-CFD Knowledge Base, as through such links the domain knowledge between industry sectors and research institutions is achieved following the process showed in Fig. 1.21.

Users can navigate through the entire knowledge base using an Internet browser. Hyperlinks provide cross-linkages between TAs, ACs, and UFRs. A

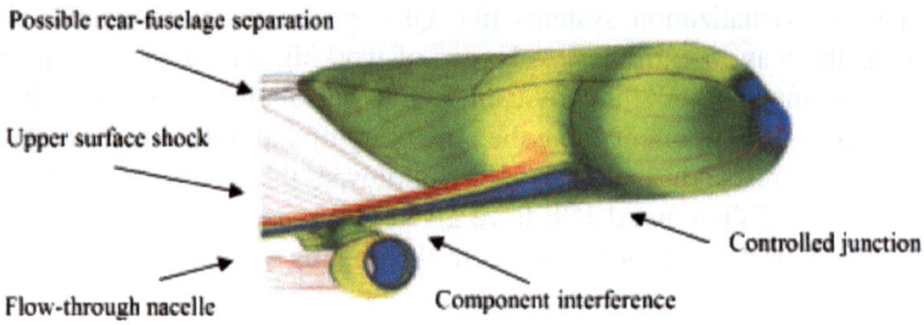

Fig. 1.19 AC example with different UFRs

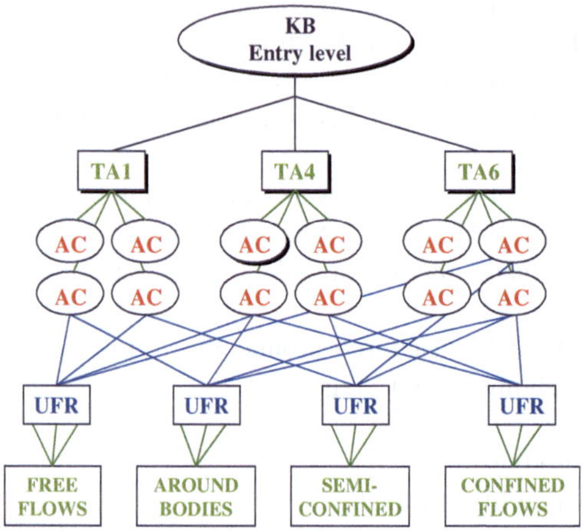

Fig. 1.20 Knowledge base structure

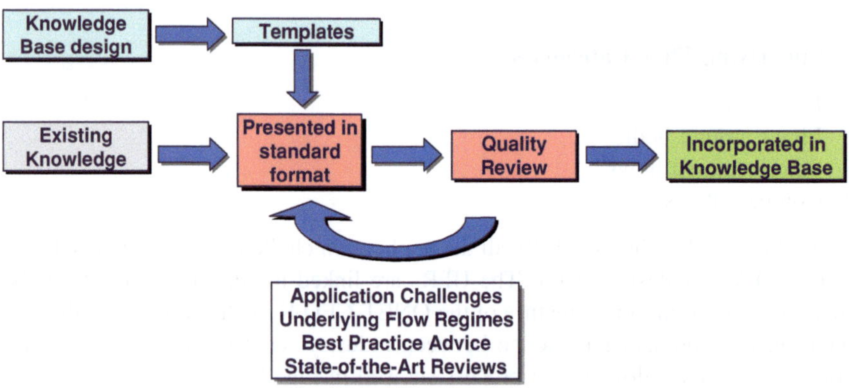

Fig. 1.21 The QNET-CFD knowledge engineering process

navigation bar on the left of the screen helps users to keep track of their location within the Knowledge Base. Users may go directly to any part of the knowledge base via content lists (Index), summary matrix tables (Matrix), and keyword search facilities.

The QNET-CFD project showed that making trusted and validated computational and experimental datasets widely available is of great importance to the EU scientific community; together with the integrated processing of CFD and EFD data, which requires appropriate access and manipulation tools. The QNET-CFD web portal has become the most effective and striking way to analyze and present the elaborated scientific data; obviously, new SV tools will emerge in the future, expected to provide more interactive visual access, manipulation, selection, and processing functionality, which basic design has been work-out in the QFView architecture defined for the ALICE project.

1.2.3 LASCOT: Visualization as Decision-Making Aid

The LASCOT project [51] was part of the EUREKA/ITEA initiative. The Information Technology European Advancement (ITEA) program for research and development in middleware is jointly promoted by the Public Authorities of the EU Members States together with large European industries, see Fig. 1.22.

The goal of LASCOT was to design, develop, and demonstrate the potential benefits of distributed collaborative decision-support technology to the "future cyber-enterprise in the global economy"; the LASCOT demonstrator was to:

- Support access to traditional information systems and to Web data;
- Enable situation assessment, and provide decision-support facilities as well as simulation and validation facilities to support business decisions;
- Include current, enhanced-as-required security tools;
- Make use of visualization technology for critical tasks such as decision making and knowledge management;
- Produce an on line learning application to facilitate the embedding of the platform, by the users.

The scenario that was retained to demonstrate the LASCOT system is illustrated in Fig. 1.23 shows the various "'actors" who are aided by the LASCOT system in monitoring a crisis situation and in making decisions. The project called for 3D visualization research and for the development of 3D presentation tools capable of treating general purpose and highly conceptual information in an appropriate and readily understandable manner. In this project, the research was focused on the visualization and manipulation of graphical content in a distributed network environment. Graphical middleware and 3D desktop prototypes [52] were specialized for situational awareness.

A state-of-the-art review during the LASCOT proposal preparation did not identify any publicly available large-scale distributed application of this kind. The

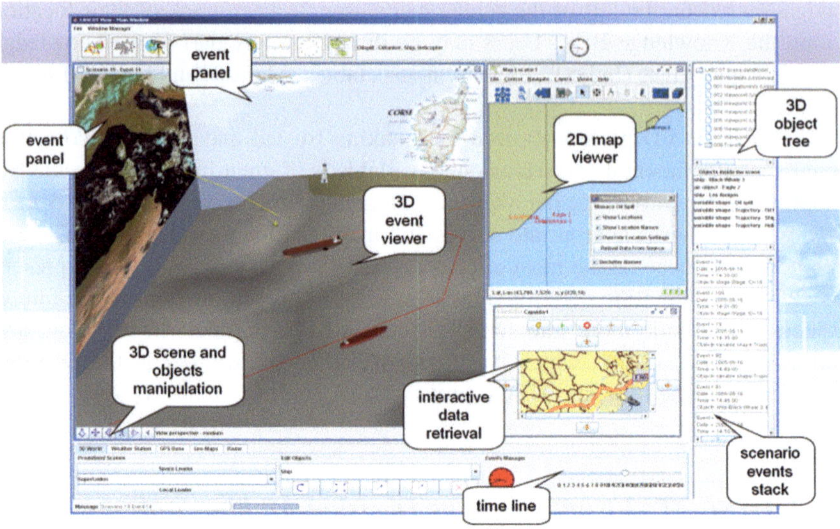

Fig. 1.22 The LASCOT application

existing proprietary solutions rely on conventional technologies and are limited to 2D representation. Our challenge was to apply the latest technologies, such as Java3D, X3D, and SOAP, compatible with average computer graphics hardware and to demonstrate a solution allowing: data flow from heterogeneous sources; interoperability across different operating systems and 3D visual representations Fig. 1.24 to enhance the end-users interaction.

We applied the Model-View-Controller (MVC) paradigm to enhance the interactivity of our 3D software components for: visualization, monitoring, and exchange of dynamic information, including spatial and time-dependent data. The software development included the integration and customization of different visualization components based on 3D Computer Graphics (Java3D) (Fig. 1.25) and Web (X3D, SOAP) technologies and applying the object-oriented approach based on Xj3D.

Cutting-edge 3D graphics technologies where integrated, including the Java-based X3D browser; we used Xj3D to visualize data from various external sources using our graphical middleware. While present software components provide highly flexible interactions and data-flows, the coupling between these components is designed to be very loose. Thus, the components can be upgraded (or even replaced) independently from each other, without loss of functionality. With SOAP messaging, intercomponents communication is completely independent of software platforms and communication transmission layers. In our approach, Java components coexist with Microsoft. NET front-end, as well as back-end implementations. This approach allows improving software development of 3D collaborative and visualization tools. Future development of an appropriate ontology could significantly improve the distributed visualization framework in SERKET.

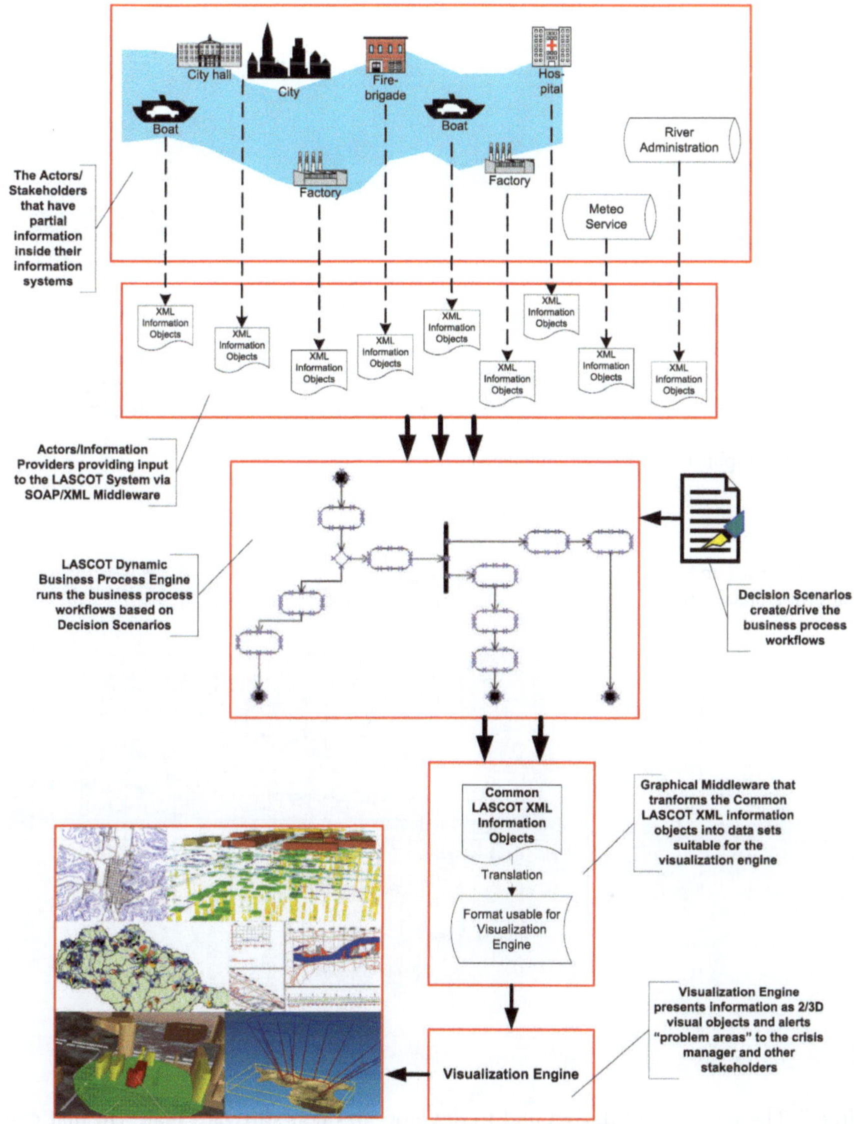

Fig. 1.23 The LASCOT scenario

1.2.4 SERKET: Security Situation Awareness

The SERKET project [53] explored a solution to the issue of security in public areas and events (Fig. 1.26), by developing an innovative system whereby dispersed data from a variety of different devices are automatically correlated, analyzed, and presented to security personnel as "the right information at the right

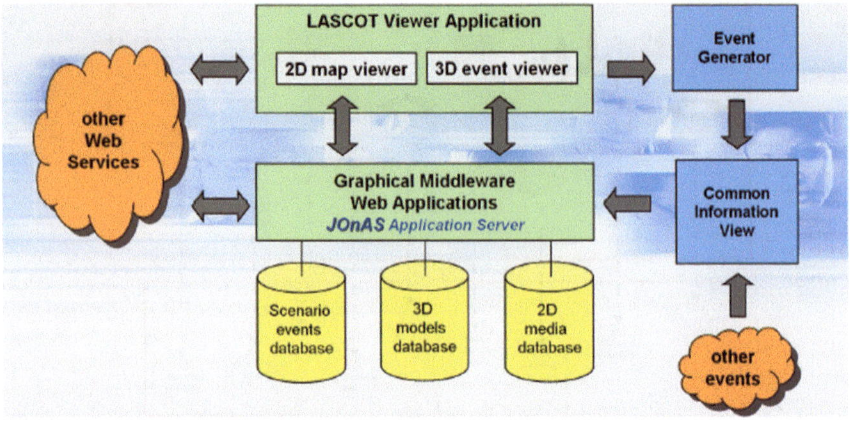

Fig. 1.24 Graphical middleware architecture

Fig. 1.25 Java and Java3D platform architecture

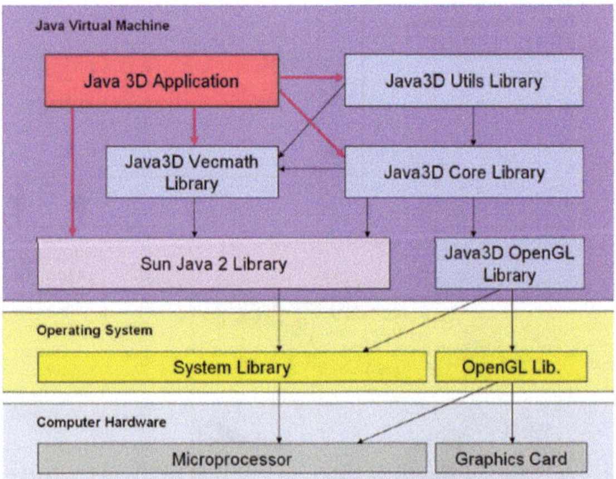

time." The aim was to design and to develop an open-software platform that can be deployed at low cost.

3D software development in SERKET is centered on the visualization and presentation engine, with special attention on the application of X3D (and XML) standards. The graphical middleware must integrate, correlate, combine, annotate, and visualize sensor data and related metadata (the application context is airport security). Using sensor data analyzed by other processing and data fusion components, the graphical middleware builds 3D scenes, which represent the objects detected by the sensors and the operational status of the sensors at their locations.

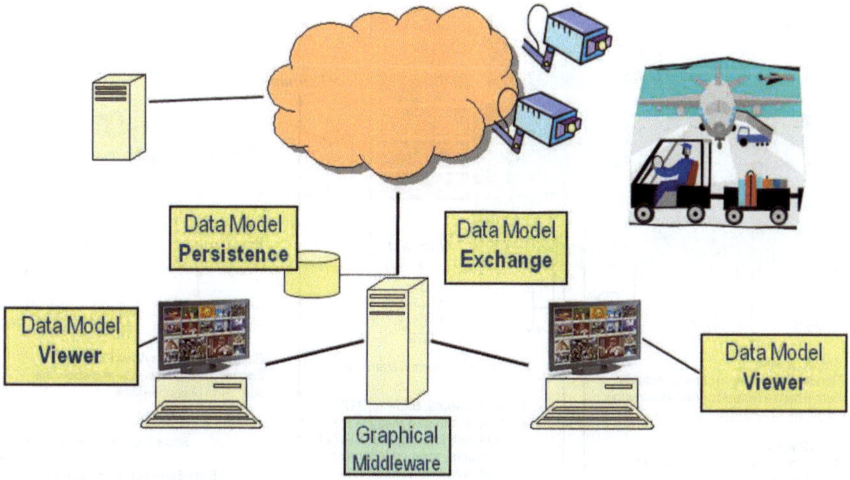

Fig. 1.26 The security SERKET scenario

Objects in the 3D scenes are annotated with metadata and/or with links to metadata describing the security context in relation to the displayed 3D objects. The 3D rendition of the situation must provide an unambiguous, highly understandable overview of the situation to the user, who should be able to switch between different levels of viewing details and select, at each level, desired viewpoints (the locations of the video cameras define the available viewpoints). The 3D model of situation-security awareness is parameterized in space and time as shown in Fig. 1.26.

The solution is based on the J2EE Application Server platform (JBoss). The interface between the components is based on SOAP messaging. 3D objects and scenes are using the X3D file format (an extended XML-based format for storing graphics). The XML markup language is used for data exchange. X3D can be extended to include metadata, thus nongraphical information related to the created graphical object. The development of the graphical middleware visualization components is done in Java, using the Eclipse environment. The 3D SERKET desktop enables access, and treatment of security data from heterogeneous sources and supports:

- 3D models with features
- Level of detail selection
- Navigation and interaction

Figure 1.27 illustrates the main functionality of the SERKET application; clearly, 3D interaction and 3D models are key elements to a new generation of visualization software which must now treat and be able to display general purpose and abstract information such as "security".

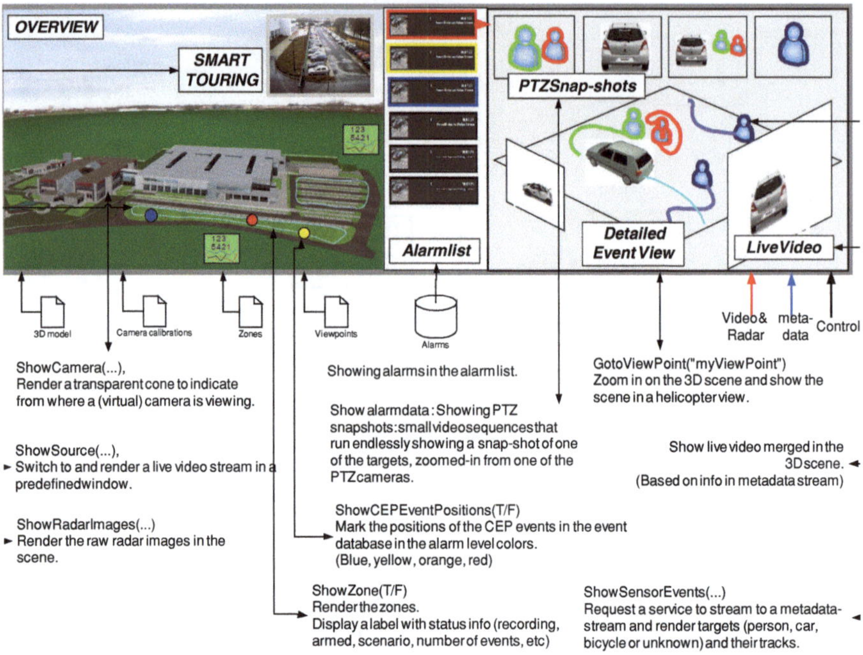

Fig. 1.27 The SERKET application

1.2.5 3D-TestBench: Multidisciplinary M&S Environment

In the 3D-Testbench project [54], the idea is to go from the "simulation-based" workflows, which have limited integration possibilities to a multidisciplinary workflows that combine heterogeneous simulation tests based the on the knowledge engineering technology. Current efforts, like SW are addressing such challenges, applying the principles of Services, for example, Web services are technologies applied in the EU-funded projects, such as SIMDAT [55].

It is expected that the 3D-TestBench can significantly change the state of the art in product development workflow, in a number of ways:

- It will enable nonspecialist stakeholders to collaborate at system level in knowledge-based engineering;
- It will reduce the number of persons with specialist skills required to design a product;
- It will empower individuals to make changes to product concepts and designs, to be innovative, and to provide confidence that the innovation is feasible.

Although in the software engineering the test-driven development (TDD) is a common practice, today, it is not applied within the multidisciplinary engineering, because there are no tools, which can support TDD in a straightforward manner

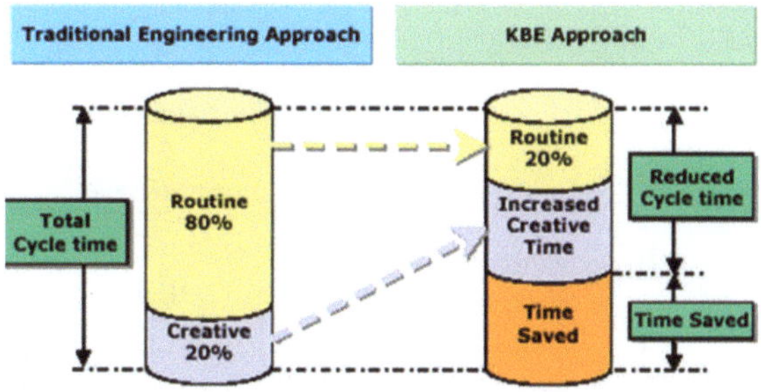

Fig. 1.28 Benefits of KBE

(Fig. 1.28). In 3D-TestBench, the identified key benefit of KBE is the accessibility of knowledge incorporated in software including its traceability and knowledge rules applied to the automated design process, which is expected to improve the modularity, and to level and ease the automation process. When M&S activities are commonly carried out, they should be effectively integrated in the development process to improve:

- ideation, innovation, understanding of complexity, and visualization of function;
- explanation of concepts, designs, operations to all stakeholders in a readily understandable manner;
- reliability: models used in simulation activities can be directly derived from the design;
- accuracy: the input parameters of a simulation can come from other simulators, or one can use several elementary simulations, each of them devoted to the modeling of a specific function;
- efficiency: integration of simulators within the same environment allows data exchange to occur automatically.

Therefore, from the standpoint of simulation end-users, the integration of multidisciplinary models addressing different aspects of system engineering is very relevant. The innovation aspects are in the integration of technologies supporting the lifecycle product management phases with the visual analytics aspects within the M&S tools, thus adding new requirements and functions which will be easily integrated in a multidisciplinary simulation and visualization environment. The applied approach within 3D-TestBench is that the complex workflows describe, both logically and practically, how the analysis of the specific virtual prototype is carried out. The XML standard is applied to describe the sequence of operations, dependencies, design parameters, and functional objectives in order to handle heterogeneity, integration of legacy systems, openness, reusability, and extendibility imposed requirements (Fig. 1.28).

For the 3D-Testbench use case (Fig. 1.29), the multidisciplinary wing design was selected. It starts with the geometry definition, which represents the reference

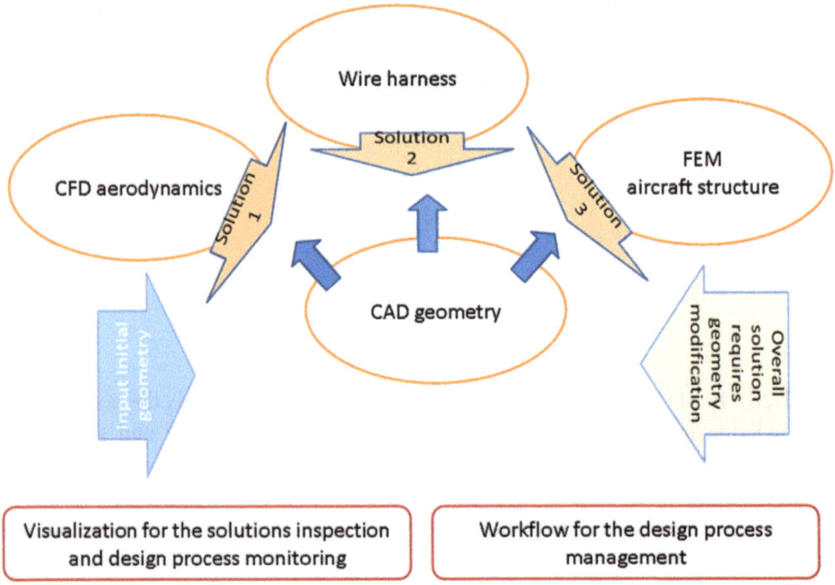

Fig. 1.29 3D-TestBench use case decomposition

point for all the selected three domain disciplines involved in the airplane wing design, thus supported with an appropriate software tools, and the following category of software tools/components have been selected:

1. CAD for geometry modeling
2. CFD for fluid flow simulations
3. FEM for strength computations
4. Wire harness design
5. Workflow manager for the design process management
6. Visualization for the solutions inspection and design process monitoring

It is important to mention that the change in the wing geometry affects the simulated results for the three selected engineering domains, thus affecting the design process, which requires an iterative path by the continuous supervision of the designer (in this case is a group of experts) covering the validation process in each of the involved discipline (structural mechanics, fluid mechanics, and wire harness), which needs to be repeated, commonly within several iterative steps before acceptable solution can be found. This process and all the supporting software tools and hardware need to constitute an integrated solution, which will be the 3D-Testbench software environment (Figs. 1.29 and 1.30).

3D-TestBench software solution, see Fig. 1.31, will enable the stakeholders to be present in the M&S design process of such complex systems, to be naturally

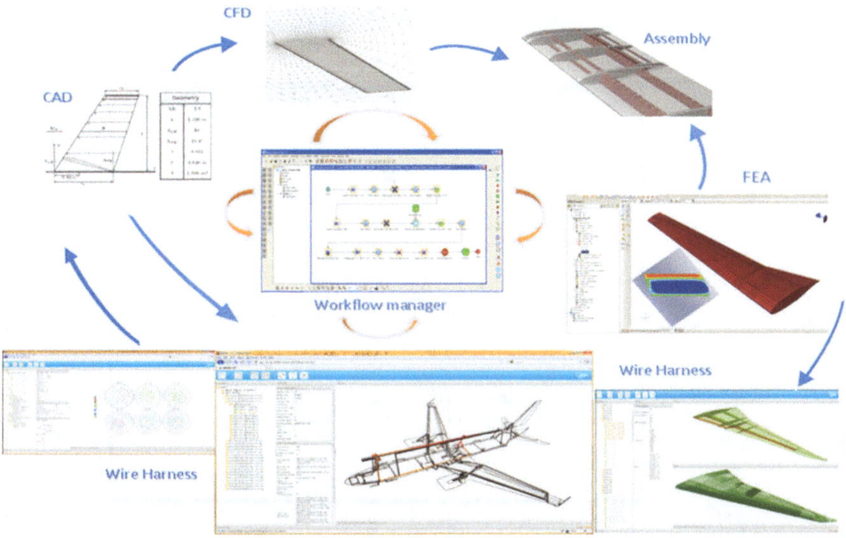

Fig. 1.30 The wing design demonstrator workflow

Fig. 1.31 3D-TestBench vision

and intuitively involved to the best of their abilities, thus collaborating with other stakeholders to: create innovative concepts, improve quality products, reduced time to market, and reduced design and development costs.

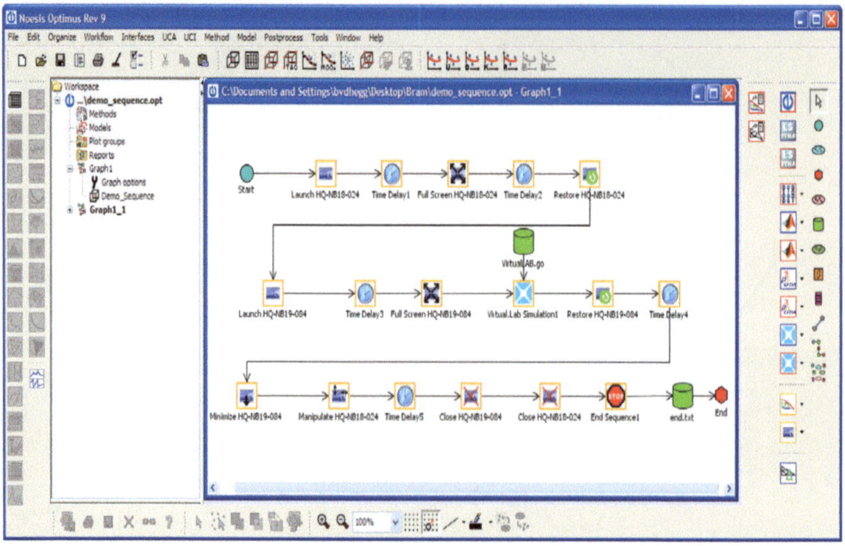

Fig. 1.32 OPTIMUS workflow management tool

Despite progress in establishing standards and in developing enabling technology, the vision of the 3D-TestBench, and its benefits, has not yet been able to be realized. The benefits of reduction of time to market, reduced development cost, improved product quality, effective (seamless and natural) collaborative multidisciplinary knowledge-based working are not yet attained. This project has integrated a number of preexisting components and developed the technology, with workflow management tool OPTIMUS (Fig. 1.32), which has been applied to enable to demonstrate the 3D-TestBench vision.

1.3 Conclusion

Today's trend in the engineering software tools is toward more intelligent multidisciplinary M&S systems, which are expected to capture the engineering intelligence in order to put in the hands of the engineers advanced tools for designing new products, which are able to perform highly complex investigations—a multiexpert team counterpart enabled through software. In order to develop such intelligent software, it is necessary to combine in addition to the Computer Graphics technologies and User Interface design, as required for SV—the missing element, the modeling of domain engineering knowledge, to clearly become part of the software development process. In this work, the advocated computer software technologies to support such multidisciplinary solutions were addressed as follows:

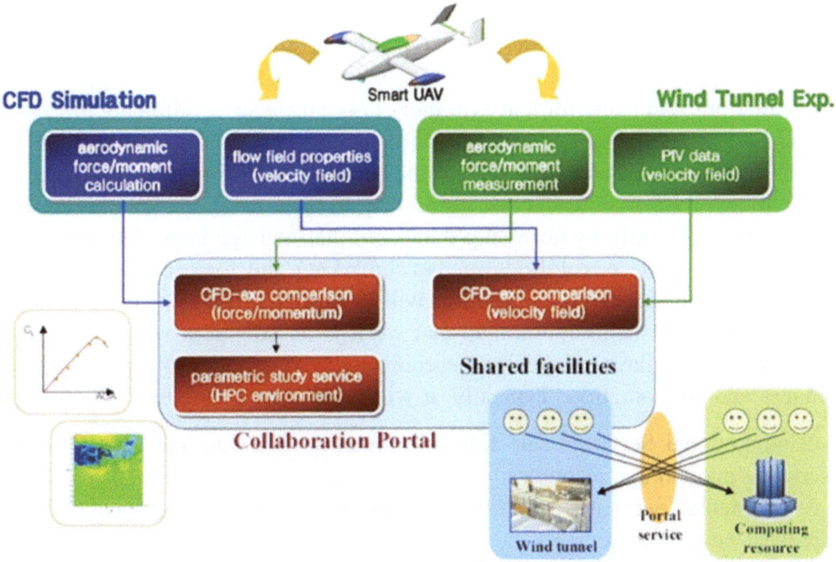

Fig. 1.33 An integrated M&S environment [57]

- M&S frameworks
- Collaborative environments
- Specialized engineering tools

Enriched with advances in:

- Visual analytics
- Ontology design
- 3D graphics
- Scientific workflows

And being developed with:

- Object-Oriented Methodology integrating
- Knowledge Engineering development tools.

The envisaged new generation software will support engineers with a homogeneous and more easily used working environment, by offering a common look-and-feel user interface to the various simulation codes and the associated datasets. In line with the given introduction example, a respective system will need to integrate the latest "best practice" software tools and will be constantly updated in order to easily run on the most-recent software/hardware platforms (Fig. 1.33). It will consist of an Internet-like portal, from which the investigator will be able to access information/knowledge/databases and processing functions, at any time and wherever the respective software is located and data stored, thus enabling accurate and efficient invocation of the engineering simulation services. It is expected that the calculations will be performed extremely fast and cheap, by allowing the use of parallel and grid computing resources [56], as the HPC and Cloud Computing R&D is promising. In order to validate the simulation results, they will be compared with experimental results and in addition, to other similar computations; this will be done efficiently by accessing databases that manage large datasets of the archived results. The possibilities for benchmarking and for exchanging knowledge and opinions between investigators will be virtually infinite. Clearly thought, a prerequisite that such environment can work is its adoption by the user community, which will have to agree on a specific codex that will enable and guarantee openness and collaboration. Typically, it will have open access to:

- Computational Services: selection of simulation software and access to processing and storage resources.
- Experimental Services: access to experimental databases with possibility to request new measurements.
- Collaborative Services: chat and video-conferencing, with usage of shared viewers (3D interactive collaboration).

The more and more evident software engineering challenge is on how to resolve the complexity and interaction between the different existing engineering models and how to process effectively the large amount of existing simulation results, which are today in the continuous grow.

On the other hand it becomes very clear that the synergy of expert's knowledge and experience, especially computer science and engineers, is of paramount importance, in order to come up with solutions, which will smoothly integrate the best practices within the engineering know-how, and transform them to the efficient and easy-to-use software solutions.

Over the last 25 years, the author has initiated and worked on many research projects with a determined focus on developing SV software and advancing the state of the art in this challenging area. As described in this text, there are still many potential avenues for future research to be followed, as addressed in the presented EU projects, which have been the necessary seeds that such R&D activities covering the multidisciplinary challenges could become reality in the engineering practice.

Acknowledgments The acknowledgment goes to all the EU projects consortia and the related participants, who took place in these challenging R&D projects span over the last two decades.

The fundings of the European Commission (EC) and the Flemish institute for Innovation and Technology (IWT) are gratefully acknowledged; the LCLMS, ALICE, LASCOT, QNET-CFD, SERKET, and 3D-TestBench projects have been instrumental in allowing carrying out such R&D work. The author is grateful to Vrije Universiteit Brussel for providing the necessary research and computer facilities for running the implementation of the engaged projects.

References

1. Vucinic, D.: Development of a Scientific Visualization System CFView—Computational Field Visualization System and Object-Oriented Software Methodology, p. 280. Lambert Academic Publishing, Saarbrücken, Germany (2010)
2. Visualization and Knowledge Discovery. Report from the DOE/ASCR Workshop on Visual Analysis and Data Exploration at Extreme Scale (2007)
3. Ma, K.-L., et al. Next-generation visualization technologies: enabling discoveries at extreme scale. SciDAC Rev. **12**, 12–21 (2009)
4. Elmagarmid, A.K., et al.: Community-cyberinfrastructure-enabled discovery in science and engineering. Comput. Sci. Eng. **10**, 46–53 (2008)
5. Morgan, K., et al.: A parallel framework for multidisciplinary aerospace engineering simulations using unstructured meshes. Int. J. Numer. Meth. Fluids **31**, 159–173 (1999)
6. Sahu, R., et al.: An object-oriented framework for multidisciplinary, multi-physics, computational mechanics. Eng. Comput. **15**, 105–125 (1999)
7. Wainer, G., Liu, Q.: Tools for Graphical specification and visualization of DEVS models. Simul. Trans. Soc. Model. Simul. Int. **85**, 131–158 (2009)
8. Muranaka, T., et al.: Development of multi-utility spacecraft charging analysis tool (MUSCAT). IEEE Trans. Plasma Sci. **36**, 2336–2349 (2008)
9. Sokolowski, J.A., Banks, C.M.: Principles of Modeling and Simulation: A Multidisciplinary Approach. John Wiley, Hoboken (2009)
10. Byungil, J.: Ultrascale collaborative visualization using a display-rich global cyberinfrastructure. IEEE Comput. Graphics Appl. **30**, 71–83 (2010)
11. Deremaux, Y.: Physically-based, real-time visualization and constraint analysis in multidisciplinary design optimization. MSc Thesis in Aeronautics and Astronautics, Massachusetts Institute of Technology (2003)
12. Brodlie, K., et al.: Visualization ontologies. Report of a Workshop held at the National e-Science Centre (2004)
13. Vucinic, D., et al.: Towards interoperable X3D models and web-based environments for engineering optimization problems (EngOpt). International Conference on Engineering Optimization Proceedings, Rio de Janeiro, Brazil (2008)
14. CEASIOM software: computerised environment for aircraft synthesis and integrated optimisation methods. http://www.ceasiom.com/ (2014)
15. Vucinic, D., et al.: CFView: an advanced interactive visualization system based on object-oriented approach. In: AIAA 30th Aerospace Sciences Meeting, Reno, Nevada (1992)
16. Ensight Software. What is EnSight. http://www.ceisoftware.com/ensight10/what-is-ensight/ (2014)
17. Duque, E., et al.: Post-processing techniques for large-scale unsteady CFD datasets. In: 45th AIAA Aerospace Sciences Meeting and Exhibit, Reno, Nevada (2007)
18. Legensky, S.M.: Recent advances in unsteady flow visualization. In: 13th AIAA Computational Fluid Dynamics Conference, Snowmass Village, CO (1997)
19. Taflin, D.E.: TECTOOLS/CFD: a graphical interface toolkit for network-based CFD. In: 36th Aerospace Sciences Meeting and Exhibit, Reno, NV (1998)
20. ParaView software. Parallel visualization application. http://www.paraview.org/paraview index.html (2014)
21. Ramachandran, P., Varoquaux, G.: Mayavi: making 3D data visualization reusable. In: 7th Python in Science Conference Proceedings, pp. 51–56, Pasadena, CA USA (2008)
22. Whitlock, B.J.: Visualization with VisIt. (UCRL-PRES-209591). Lawrence Livermore National Laboratory (2005)
23. Bertini, E., Lalanne, D.: Investigating and Reflecting on the Integration of Automatic Data Analysis and Visualization in Knowledge Discovery. ACM SIGKDD Explorations **11**(2), 9–18 (2009)

24. Keim, D., Mansmann, F., Oelke, D., Ziegler, H.: Visual Analytics: Combining Automated Discovery with Interactive Visualizations, International Conference on Discovery Science, Lecture Notes in Computer Science (LNCS). Discovery Science, vol. 5255, pp. 2-14, Springer-Verlag, Budapest (2008)
25. Zhanjun, L., et al.: A Methodology for Engineering Ontology Acquisition and Validation, vol. 23, pp. 37–51. Cambridge University Press, New York (2009)
26. Extensible 3D (X3D) Specification, ISO/IEC 19775-1:2013. http://www.web3d.org/files/specifications/19775-1/V3.3/index.html (2013)
27. Craig, A., et al. (2006) Evaluating X3D for use in software visualization. In: Proceedings of the ACM Symposium on Software Visualization, Brighton, United Kingdom
28. Geroimenko, V., Chen, C.: Visualizing Information Using SVG and X3D: XML-Based Technologies for the XML-Based Web. Springer, London (2005)
29. XSL Transformations (XSLT) Version 1.0 W3C Recommendation. http://www.w3.org/TR/xslt (1999)
30. Lassila, O., Swick, R.R. (eds.): Resource Description Framework (RDF), Model and Syntax Specification, W3C Recommendation (1999)
31. Zuo, Z., Zhou, M.: Web ontology language OWL and its description logic foundation. In: Proceedings of the 4th International Conference on Parallel and Distributed Computing, Applications and Technologies, PDCAT, pp. 157–160 (2003)
32. Vucinic, D.: Object oriented programming for computer graphics and flow visualization. In: VKI Lecture Series on Computer Graphics and Flow Visualization in CFD, ed. von Karman Institute for Fluid Dynamics, Brussels, Belgium, vol. 7, p. 37 (1991)
33. Vucinic, D.: Development of a scientific visualization system. PhD Thesis, Department of Mechanical Engineering, Vrije Universiteit Brussel (2007)
34. Yolanda, G.: From Data to Knowledge to Discoveries: Artificial Intelligence and Scientific Workflows, vol. 17, pp. 231–246. IOS Press, Amsterdam (2009)
35. Vouk, M.A., Altintas, I., Barreto, R., Blondin, J., Cheng, Z., Critchlow, T., Khan, A., Klasky, S., Ligon, J., Ludaescher, B., Mouallem, P.A., Parker, S., Podhorszki, N., Shoshani, A., Silva, C. (eds.): Automation of Network-Based Scientific Workflows (IFIP International Federation for Information Processing Grid-Based Solving Environments). Springer, Boston (2007)
36. Doran, P., et al.: Ontology module extraction for ontology reuse: an ontology engineering perspective. In: Proceedings of the 16th ACM Conference on Information and Knowledge Management, Lisbon, Portugal (2007)
37. Barseghian, D., et al.: Workflows and extensions to the Kepler scientific workflow system to support environmental sensor data access and analysis. Ecological Informatics **5**, 42–50 (2010)
38. Tohline, J.E., et al.: A Customized Python Module for CFD Flow Analysis within VisTrails. Comput. Sci. Eng. **11**, 68–72 (2009)
39. Vucinic, D., Hirsch, C.: Computational flow visualization system at VUB (CFView 1.0). VKI Lecture Series on Computer Graphics and Flow Visualization in CFD, 1989-07, Brussels, Belgium (1989)
40. Désidéri, J.-A., et al.: Hypersonic Flows for Reentry Problems: Survey Lectures and Test cases for Analysis, vol. 1, pp. 22–25. Springer, Heidelberg (1990)
41. Torreele, J., Keymeulen, D., Vucinic, D., van den Berghe, C.S., Graat, J., Hirsch, Ch.: Parallel CFView : a SIMD/MIMD CFD visualisation system in a heterogeous and distributed environment. In: International Conference on Massively Parallel Processing, Delft, The Netherlands (1994)
42. Lang, U.: A software environment for cooperative simulation and visualization in the aerospace field. High Perform. Comput. Networking Lect. Notes Comput. Sci. **797**, 70–76 (1994)
43. Jalby, W.: Europe: building confidence in parallel HPC. Comput. Sci. Eng. (1994). doi: 10.1109/MCSE.1994.10034

44. Grijspeerdt, K., Backx, E., Rammant, J.P.: LCLMS, an advanced database environment for the development of multimedia courses. In: Computers in the practice of building and civil engineering, Worldwide ECCE symposium, Finland (1997)
45. Vucinic, D., et al.: QFView—an internet based archiving and visualization system. In: 39th Aerospace Sciences Meeting and Exhibit, Reno, Nevada (2001)
46. Vucinic, D., et al.: Fast and convenient access to fluid dynamics data via the World Wide Web. European Congress on Computational Methods in Applied Sciences and Engineering (ECCOMAS), Invited Technological Session on Parallel Multidisciplinary Simulation Environments, Barcelona, Spain (2000)
47. Shannon, B.: Java 2 Platform, Enterprise Edition: Platform and Component Specifications. Addison-Wesley, Boston (2000)
48. Purba, S.: High-Performance Web Databases: Design, Development and Deployment. Auerbach, Boca Raton (2001)
49. Eberhart, A., Fischer, S.: Java tools: Using XML, EJB, CORBA, Servlets and SOAP. Wiley, New York (2002)
50. QNET-consortium, GTC1-CT99-10030 QNET-CFD. Final Technical Report (2004)
51. LASCOT project info. http://www.bull.com/lascot/index.html (2014)
52. Vucinic, D., et al.: Distributed 3D information visualization, towards integration of the dynamic 3D graphics and Web services. In: 1st International Conference on Computer Graphics Theory and Applications, Setúbal, Portugal (2006)
53. SERKET project info. http://www.multitel.be/image/research-development/research-projects/serket.php (2006)
54. ITEA2_3D-TestBench_consortium, 3D-TestBench. Full Project Proposal Revision (2007)
55. SIMDAT FP6 Grid project, data grids for process and product development using numerical simulation and knowledge discovery. ftp://ftp.cordis.europa.eu/pub/ist/docs/grids/simdat_fact_sheet.pdf (2008)
56. Pandey, S., et al.: A grid workflow environment for brain imaging analysis on distributed systems. Concurrency and Comput. Pract. Experience **21**, 2118–2139 (2009)
57. Jeong, M.-J., et al.: e-AIRS: aerospace integrated research systems. In: International Symposium on Collaborative Technologies and Systems (CTS'07), Orlando, Florida, USA (2007)

Chapter 2
A Distributed Architecture for Simulation Environments Based on Game Engine Systems

Mark Joselli, Marcelo Zamith, Luis Valente, Bruno Feijó, Fabiana R. Leta and Esteban Clua

Abstract Simulation systems are becoming common in different knowledge fields, such as aeronautics, defense, and industrial applications, among many others. While in the past these systems where mostly based on typical Virtual Reality Environments, with the advance of the game industry simulators are being developed using typical game engines and gaming software architectures. Distributed computing is being used in several fields to solve many computation intensive problems. Due the complexity of Simulation systems, this architecture can also be used, devoting host processing to renderization, which is usually the task that simulators spend most of its processing time. By using distributed computing, simulators could need softer system requirements, since the main loop would be distributed. This work presents concepts of simulator software, which is based on the main loop technique. After describing state-of-the-art concepts, we present an efficient automatic load balancing and distributing logic computation among several computers for simulators.

M. Joselli · M. Zamith · L. Valente · E. Clua (✉)
Computing Institute, Universidade Federal Fluminense-UFF, Niterói, RJ, Brazil
e-mail: esteban@ic.uff.br

M. Joselli
e-mail: mjoselli@ic.uff.br

M. Zamith
e-mail: mzamith@ic.uff.br

L. Valente
e-mail: lvalente@inf.puc-rio.br

B. Feijó
Informatic Department, PUC-Rio, ICAD Games, Rio de Janeiro, RJ, Brazil
e-mail: bfeijo@inf.puc-rio.br

F. R. Leta
Mechanical Engineering Department, Universidade Federal Fluminense-UFF, Niterói, RJ, Brazil
e-mail: fabianaleta@id.uff.br

F. R. Leta (ed.), *Visual Computing*, Augmented Vision and Reality 4,
DOI: 10.1007/978-3-642-55131-4_2, © Springer-Verlag Berlin Heidelberg 2014

Keywords Simulation · Real-time visualization · Game engine architecture · Task distribution

2.1 Introduction

Increasing realism level in virtual simulations depends not only on the enhancement of modeling and rendering effects, but also on the improvement of different aspects such as animation, artificial intelligence of the characters, and physics simulation. Real-time systems are defined as solutions that have time constraints to run their tasks. Hence, if the system is unable to execute its work under some time threshold, it will fail. In order to achieve such constraints, the main loops have to be carefully implemented. The main loop is the design pattern of such kind of applications.

Real-time simulators are applications that employ knowledge of many different fields, such as computer graphics, artificial intelligence, physics, computer networks, and others. While these are typical requirements found in games, simulations usually require these features with much more accuracy. More, computer simulators are also interactive applications that exhibit three general classes of tasks: data acquisition, data processing, and data presentation. Data acquisition is related to gathering data from input devices as keyboards, mice, and dedicated interfaces, depending on the simulator. Data processing tasks consists on applying logic rules, responding to user commands, simulating physics, and artificial intelligence behaviors. Data presentation tasks relate to providing feedback to the user about the current simulation state, usually through images and audio. Many simulators are included in multiuser environments, requiring the usage of distribution and logical partitioning of the scene [1].

Simulators are interactive real-time systems and have time constraints to execute all of their processes and to present the results to the user. If the system is unable to do its work in real time, it will lose its interactivity and consequently it will fail. A common parameter for measuring the performance is frames per second (FPS). The general lower acceptable bound for a game is 16 FPS. There are no higher bounds for a FPS measurements, but when the refresh rate of the video output (a computer monitor) is inferior to the game application refresh rate, some generated frames will not be presented to the user (they will be lost). One motivation for designing loops optimizations is to better achieve an optimal FPS rate for the application. Doing so, it is possible to spend more time with higher precision physical calculation or more complex logic behaviors.

The architecture that we present in this paper follows a similar concept as cloud and distributed computing, where machines across the Internet shares resources, software, and information, where the user's computer can use other resources available on the network, to help it in processing the application. By using this approach, a computer with less computing power could join the simulation session, by relaying the effort to process the system to the network cloud.

This work summarizes several works of the authors, more specially [2–6], which presented different approaches for automatic task distribution between CPU cores and GPU in game and simulation environments, for single or multithreaded loops. In this work, we extend and describe more details related to the distribution of tasks of main loop and a smart load/balancing of the tasks.

This chapter is organized as follows: Sect. 2.2 presents a set of real-time loop model concepts found in the literature, specially coming from the gaming field. Section 2.3 presents a framework architecture concept that can be used for multiplatform simulation environments, with dynamic load balancing. Finally we present the conclusions.

2.2 Related Works

The real-time loop is the underlying structure games and real-time simulations are built upon. These loops are regarded as real time due to the time constraints to run the game-related tasks. This loop may become a bottleneck in complex simulations, due the complexity of physics and visualization requirements.

As mentioned earlier, the tasks that a computer simulation should execute can be broken down into three general groups: data acquisition, data processing, and presentation. Data acquisition means gathering data from available input devices, such as, mice, joysticks, keyboards, motion sensors, and dedicated input devices, such as an aircraft panel or car instrumentation. The data processing part refers to applying the user input into the application (user commands), applying simulation rules (the simulation logic), simulating the world physics, the artificial intelligence, and related tasks. The presentation refers to providing feedback to the user about the current simulation state, through images, audio, and in some cases motion.

Real-time simulators provide the illusion that everything is happening at once. Since these systems are interactive applications, the user will not have a good experience if the systems are not able to deliver its work on time. This issue characterizes these systems as heavy real-time applications. Although the real-time loop represents the heart of real-time simulations, it is not easy to find academic works specifically devoted to this subject. The works by Valente et al. [7], Dalmau [8], Dickinson [9], Watte [10], Gabb and Lake [11], and Monkkonen [12] are among the few ones.

The simplest real-time loop models are the coupled ones. The Simple Coupled Model [13] is perhaps the most straightforward approach to modeling real-time loops. It consists of sequentially arranging the tasks in a main loop. Figure 2.1 depicts this model.

The uncoupled models separate the rendering and update stages, so they can run independently, in theory. These models consider single-thread [9, 13] and multithread designs [11–13]. The Multithread Uncoupled Model [13] and the Single-thread Uncoupled Model [13] try to bring determinism to the simulator execution

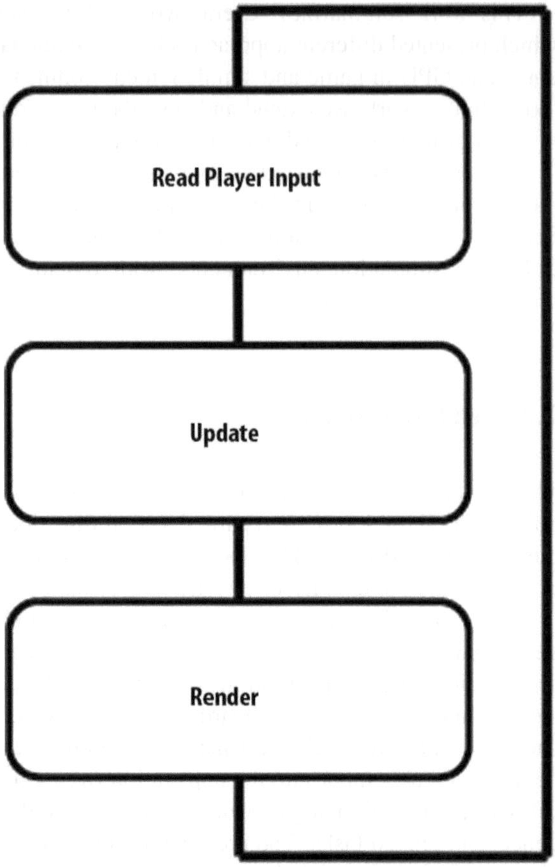

Fig. 2.1 Simple coupled model

by feeding the update stage with a time parameter. Figures 2.2 and 2.3 illustrate these models, respectively.

By using these models, the application has a chance to adjust its execution with time, so the system can run the same way in different machines. More powerful machines will be able to run the simulation more smoothly, while less powerful ones will still be able to provide some experience to the user. This requirement is also important due the fact that hardware configuration change fast, while the simulator may have a longer time of life.

Although these are working solutions, time measuring may vary greatly in different machines due to many reasons (such as process load), making it difficult do reproduce it faithfully. For example, some training sessions may require a scene replay feature [5], which may not be trivial to implement if it is not possible to run some part of the loop sequence in a deterministic way. Other features as network module implementation and program debugging [5] may be easier to implement if the loop uses a deterministic model. Another issue is that running some simulations too frequently, like AI and the game logic, may not yield better results.

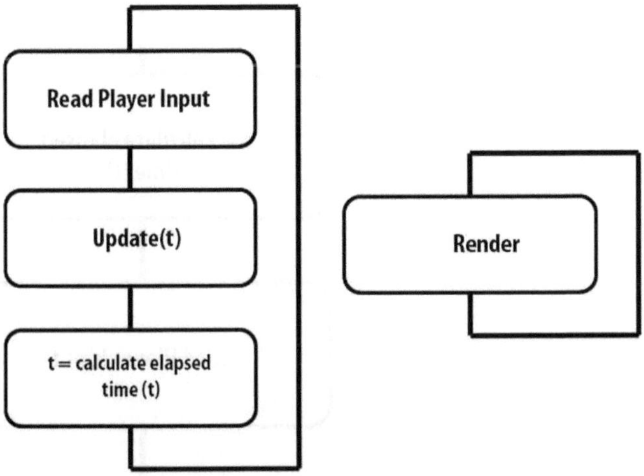

Fig. 2.2 Multithread uncoupled model

Hence, the models proposed in [9, 13, 14] try to address these issues. The Fixed-Frequency Uncoupled Model outlined in [13] features another update stage that runs at a fixed frequency, besides the time-based one. Dalmau [8] presents a similar model, although not naming it explicitly. These works describe the model using a single-thread approach. Figure 2.4 illustrates the Fixed-Frequency Uncoupled Model.

The model described in [9] presents just one update stage that runs at a fixed-frequency, whose main objective is to attain reproducibility. Another interesting model is the one used in the Microsoft XNA framework [14]. The XNA model has an update stage that runs at a fixed frequency or freely, but not both. The user is able to set a parameter that informs the XNA framework about which one to use. While this is an old framework, modern engines still keep using the same approach. This is especially relevant for simulators, since they typically use high profile machines.

Dealing with concurrent programming introduces another set of problems, such as data sharing, data synchronization, and deadlocks. Also, as Gabb and Lake [11] states, that not all tasks can be fully parallelized due to dependencies among them. As examples, a dynamic simulation element cannot move until the complete logic is computed, and visualization cannot be performed until the game state is updated. Hence, serial tasks represent a bottleneck to parallelizing simulation computation.

Monkkonen [12] presents models regarding multithread architectures that are grouped into two categories: function parallel models and data parallel models. The first category is devoted to models that present concurrent tasks, while the second one tries to find data that can be processed entirely in parallel. The Synchronous Function Parallel Model [12] proposes to allocate a thread to all tasks that are (theoretically) independent of each other. For example, performing complex physics simulation while calculating animation. Figure 2.5 illustrates this model.

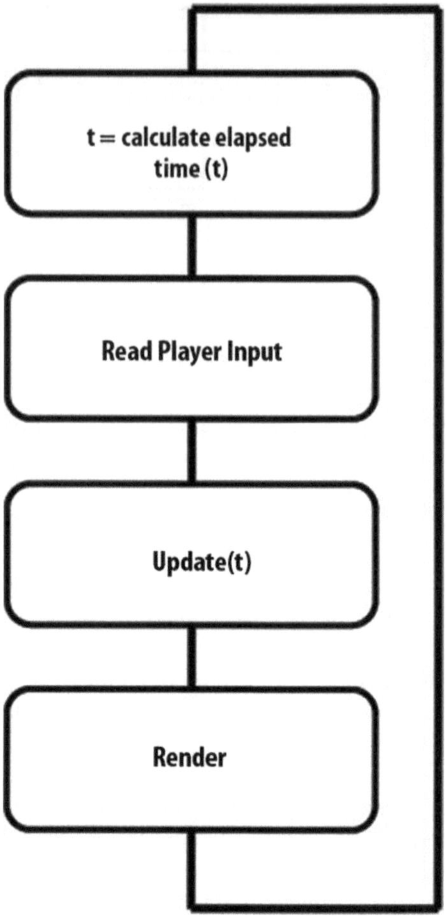

Fig. 2.3 Single-thread uncoupled model

Monkkonen states that this model is limited by the amount of available processing cores, and the parallel task should have little dependency on each other. The Asynchronous Function Parallel Model [12] is the formalization of the idea found in [11]. This model does not present a main loop. Figure 2.6 illustrates the model.

Different threads run the simulation tasks by themselves. The model is categorized as asynchronous because the tasks do not wait for the completion of other ones to perform their job. Instead, the tasks use the latest computed result to continue processing. For example, the visualization task would use the latest completed physics information to draw the objects. This measure decreases the dependency among tasks. However, task execution should be carefully scheduled for this scheme to work nicely. Unfortunately, this is often out of the scope of the application. Also, serial parts of the application (like rendering) may limit the performance of parallel tasks [11].

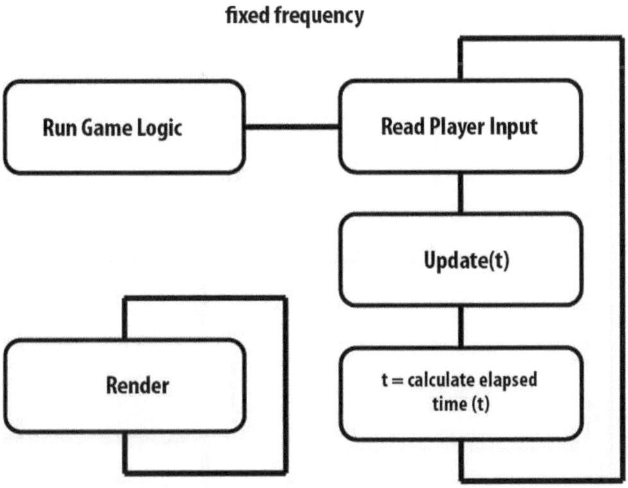

Fig. 2.4 Fixed-frequency uncoupled model

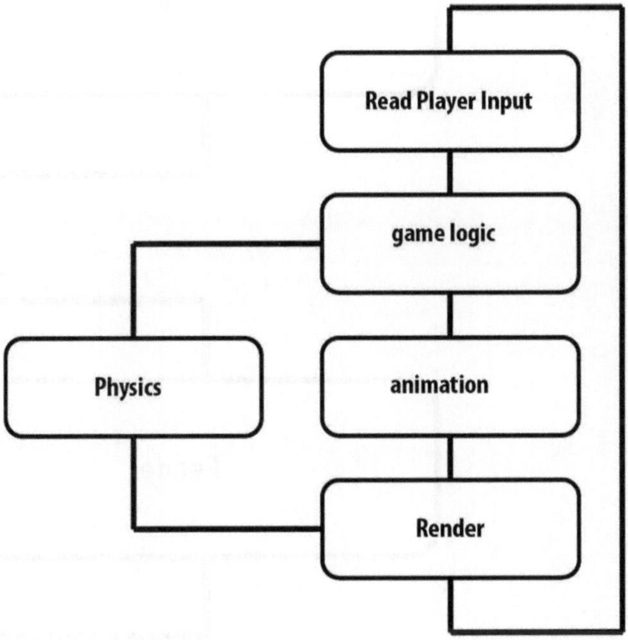

Fig. 2.5 Synchronous function parallel model

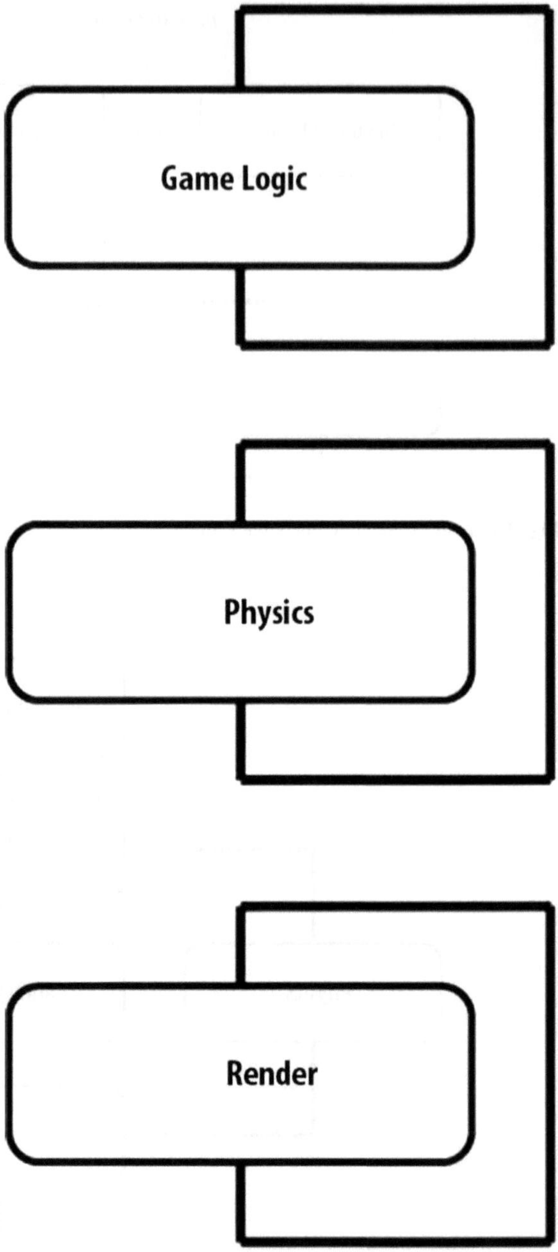

Fig. 2.6 Asynchronous function parallel model

Rhalibi et al. [15] show a different approach for real-time loops that is modeled by taking the tasks and its dependency into consideration. It divides the loop steps in three concurrent threads, creating a cyclic-dependency graph, to organize the

Fig. 2.7 Data parallel model

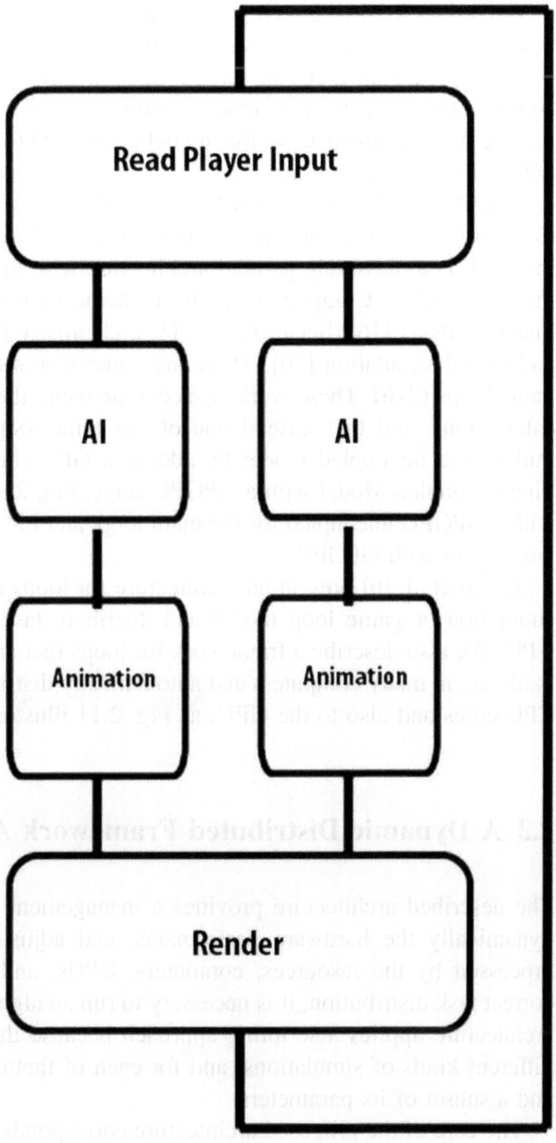

task ordering. In each thread, the tasks for rendering and update are divided taking into consideration their dependency.

The Data Parallel Model [12] uses a different paradigm where data are grouped in parallel sections of the application where they are processed. So, instead of using a main loop with concurrent parts that process all data, the Data Parallel Model proposes to use separate threads for sets of data (like game objects). This way, the objects run their own tasks (like AI and animation) in parallel. Figure 2.7 depicts this approach.

According to Monkkonen [12], this model scales well because it can allocate as many processing cores as they are available. Performance is limited by the amount of data processing that can run in parallel. An important issue is how to synchronize communication of objects running in different threads. The author states that the biggest drawback of this model is the need to having components designed with data parallelism in mind.

GPGPU stands for General Purpose GPU Computing. The focus on GPGPU Computing has been increasing since graphics hardware had become programmable. It is a massively parallel architecture with more powerful processing than the CPUs. GPU Computing has been theme of research on diverse areas like: image analysis [16], linear algebra [17], chemistry [18], physics simulation [19], and crowd simulation [20]. There are some works that discuss using GPU with game loops [2–6]. These works concentrate using the GPU mostly for the physics calculations, and they extend one of the game loops presented previously, i.e., multithread uncoupled model by adding a GPU stage. Figure 2.8 illustrates the Single Coupled Model with a GPGPU stage, Fig. 2.9 presents Multithread Model with a GPGPU uncoupled for the main loop, and Fig. 2.10 depicts the Multithread Uncoupled with GPGPU.

Joselli et al. [6] present an architecture for loops that is able to implement any simulation or game loop model and distribute tasks between the CPU and the GPU. We also describe a framework for loops that are able to detect the available hardware in many computers and automatically distribute tasks among the various CPU cores and also to the GPU, as Fig. 2.11 illustrates.

2.3 A Dynamic Distributed Framework Architecture

The described architecture provides a management layer that is able to analyze dynamically the hardware performance and adjust the amount of tasks to be processed by the resources, computers, CPUs, and GPUs. In order to make a correct task distribution, it is necessary to run an algorithm for the estimation. The architecture applies a scripting approach because the loop can be used in many different kinds of simulations, and for each of them it uses a different algorithm and a subset of its parameters.

The core of the proposed architecture corresponds to the Task Manager and the Hardware Check class. The Task Manager schedules tasks in threads and changes which processor handles them whenever it is necessary. The Hardware Check detects the available hardware configuration capabilities.

Additionally, with this architecture one can implement any loop model previously presented in this work. Also the heuristics presented in [4] can be adapted for this framework. An earlier version of this architecture was first presented in [6] and it is based on the concept of tasks. A task corresponds to some work that the application should execute, for instance: reading player input, rendering, and updating application objects.

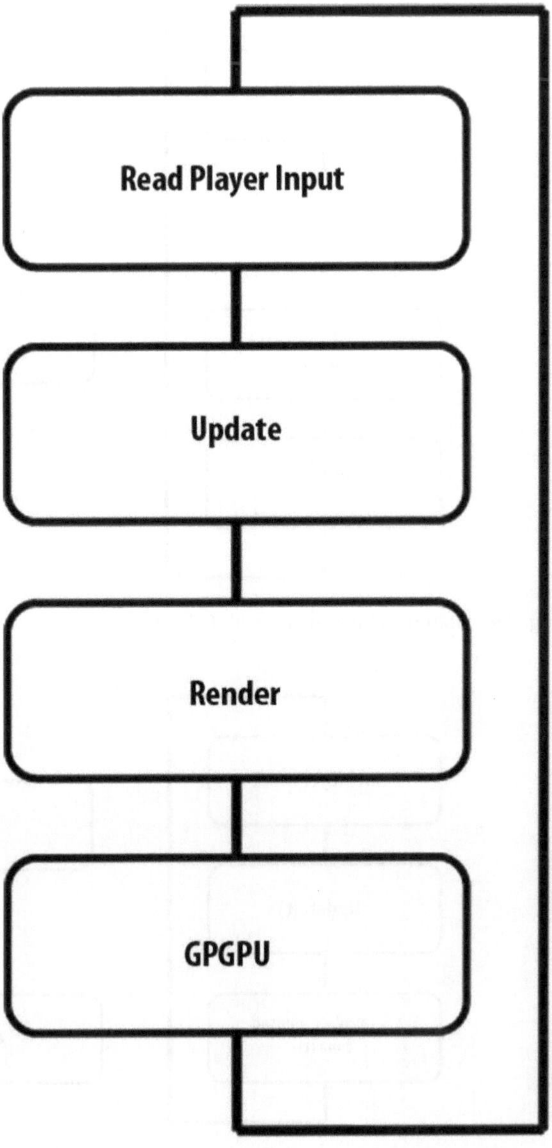

Fig. 2.8 Single coupled model with a GPGPU stage

In this architecture, a task can be anything that the application should work toward processing. However, not all tasks can be processed by all processors. Usually, the application defines three groups of tasks. The first one consists of tasks that can be modeled only for running on the CPU, like reading player input, file handling, and managing other tasks. The second group consists of tasks that run in the GPU, like the presentation of the scene. The third group can also be modeled for running on both processors and also for distributing among

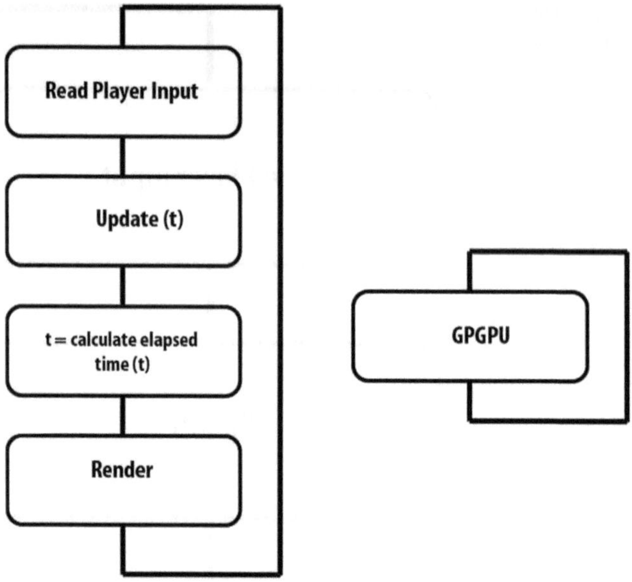

Fig. 2.9 Multithread uncoupled with GPGPU

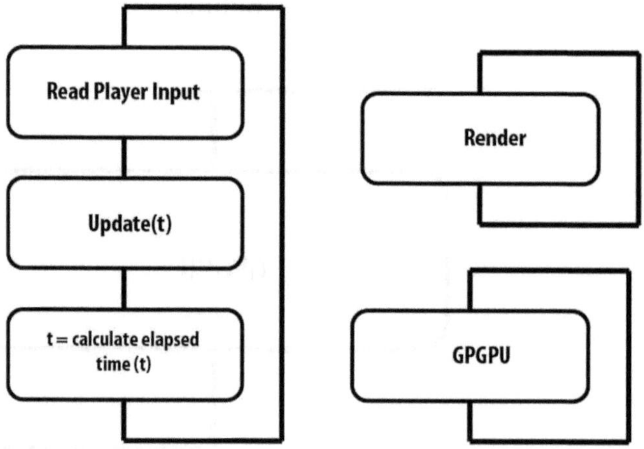

Fig. 2.10 Multithread with render uncoupled and with GPGPU stage

computers. These tasks are responsible for updating the state of some objects that belongs to the application, like AI and Physics.

The Task class is the virtual base class and has six subclasses: Input Task, Update Task, Presentation Task, Hardware Check Task, Network Check Task, and Task Manager (Fig. 2.12). The first three are also abstract classes. The fourth and fifth is a special class to check the hardware and network connection speed. The

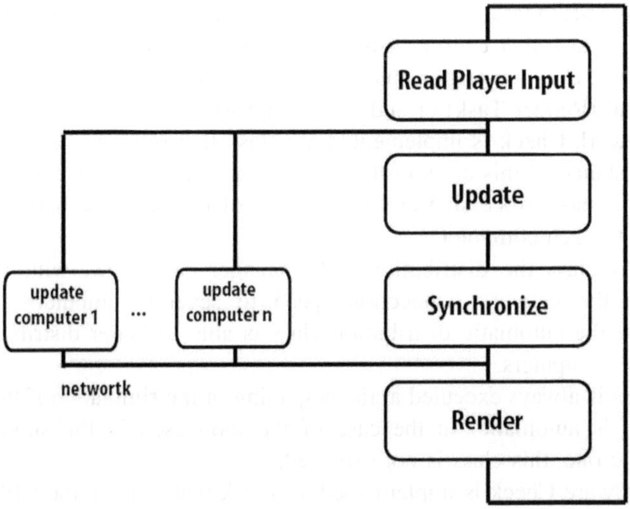

Fig. 2.11 The distributed system loop

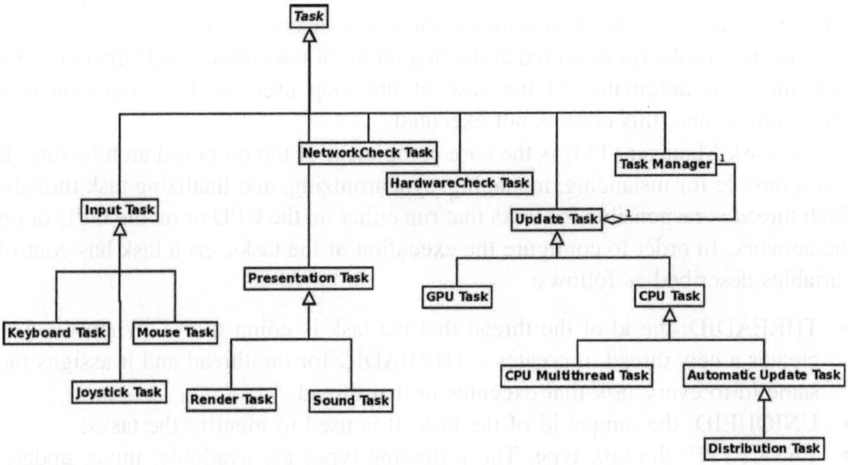

Fig. 2.12 Framework UML diagram

latter is a special class whose work consists on performing the distribution of tasks. This special class is used by the Automatic Update Task, which distributes tasks between CPU cores and GPU, and Distribution Task, which distributes tasks among computers.

The Input Task classes and subclasses handle user input-related issues. The Update Task classes and subclasses are responsible for updating the loop state. The CPU Update class should be used for tasks that run on the CPU, the GPU Update class corresponds to tasks that run on the GPU, and the CPU Multithread task class correspond to task that can be distributed among CPUs cores. The Presentation Task and subclasses are responsible for presenting information to the user, which can be visual (Render Task) or audio (Sound Task).

The Network Check is implemented as a task that runs on the CPU. There is only one instance of this class in the application. This class checks the available computers for task processing and keeps track of the available bandwidth of the connection to each computer.

With this class the distribution task manager can know, without previous knowledge, the network connection speed to several computers. Using this information, the automatic distribution class is able to better distribute the task between the computers.

This class is always executed at the beginning of the simulation if the real-time loop model is automatic. In the case of the loop used in the simulation is a deterministic one, this class is not executed.

The Hardware Check is implemented as a task that runs on the CPU. There is only one instance of this class in the application. This class checks the available hardware and keeps track of the configuration of each computer, i.e., the number of CPU cores and GPUs (with their capabilities) available in the system.

With this class the automatic task manager can know, without previous knowledge, the available hardware for the end user computer.

This class is always executed at the beginning of the simulation if the real-time loop model is automatic. In the case of the loop used in the simulation is a deterministic one, this class is not executed.

The Task Manager (TM) is the core component of the proposed architecture. It is responsible for instancing, managing, synchronizing, and finalizing task threads. Each thread is responsible for tasks that run either on the CPU or on the GPU or on the network. In order to configure the execution of the tasks, each task has control variables described as follows:

- THREADID: the id of the thread that the task is going to use. When the TM creates a new thread, it creates a THREADID for the thread and it assigns the same id to every task that executes in that thread;
- UNIQUEID: the unique id of the task. It is used to identify the tasks;
- TASKTYPE: the task type. The following types are available: input, update, presentation, and manage;
- DEPENDENCY: a list of the tasks (ids) that this task depends on to execute.

With that information, the TM creates the task and configures how the task is going to execute. A task manager can also hold another task manager, so it can use it to manage some distinct group of tasks. An example of this case is the automatic update tasks and the distribution task.

The Task Manager acts as a server and the tasks act as its clients, as every time a task ends, it sends a message to the Task Manager. The Task manager then checks which task it should execute in the thread.

When the Task Manager uses a multithread loop model, it is necessary to apply a parallel programming in order to identify the shared and nonshared sections of the application, because they should be treated differently. The independent sections compose tasks that are processed in parallel, like the rendering task. The shared sections, like the update tasks, need to be synchronized in order to guarantee mutual-exclusive access to shared data and to preserve task execution ordering.

Although the threads run independently from each other, it is necessary to ensure the execution order of some tasks that have processing dependence. The architecture accomplishes this by using the DEPENDENCY variable list that the Task Manager checks to know the task execution ordering.

The processing dependence of shared objects needs to use a synchronization object, as applications that use many threads do. Multithread programming is a complex subject, because the tasks in the application run alternately or simultaneously, but not linearly. Hence, synchronization objects are tools for handling task dependence and execution ordering. This measure should also be carefully applied in order to avoid thread starvation and deadlocks. The TM uses semaphores as the synchronization object.

2.3.1 An Automatic and Dynamic Update Task

This is an important module that we propose. The purpose of this class is to define which processor will run the task. The class may change the task's processor during the application execution, which characterizes a dynamic distribution.

One of the major features of this new architecture is to allow dynamic and automatic task allocation between the CPU threads and GPU. In order to do that it uses the Automatic Update Task class. This task can be configured in order to be executed in five modes: one CPU thread only, multithread CPU, GPU only, in the automatic distribution between the hardware detected by Hardware Check class.

In order to execute on the multithread CPU mode, there are some requirements: a parallel CPU implementation must be provided for the CPU; for executing on the GPU mode a GPU implementation must be provided; and in order to make use of the automatic distribution all the implementations must be provided accordingly to the mode. The distribution is done by a heuristic in a script file. Also a configuration on how the heuristic is going to behave is needed, and for that a script configuration file is required. The script files can be implemented in any scripted language and in our work we developed using Lua [21].

The Automatic Update Task acts like a server and its tasks as clients. The role of the automatic update task is to execute a heuristic to automatic determine in which processor the task will be executed. The Automatic update task executes the heuristic and determines which client will execute the next task and will send a message to the chosen client, allowing it to execute. Also, every time the clients finish a task they send a message to the server to let it know it has finished. Figure 2.13 illustrates this process.

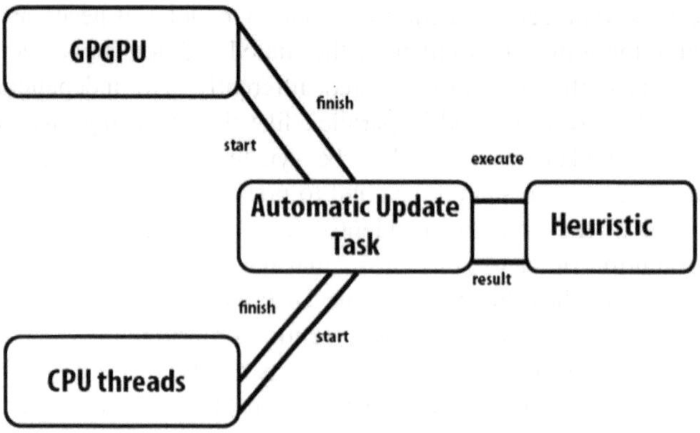

Fig. 2.13 The automatic update task class and messages

One of the main features of the proposed architecture is scheduling a task to run on another processor (CPU core to GPU or GPU to CPU core or CPU core to other CPU core) during its execution. In these cases, the task state is pushed to the task's own stack (and later restored) regardless of the processor type. For example, in time t1 the GPU processes a physics task and in time t_2 this task is scheduled to the CPU. When the task starts to run again (now in the CPU), the Task Manager reloads the task state from the tasks stack and signals it that the processor type has changed. The task priority is changed to a value of zero, which means that the task is placed on the front of the task queue. This measure is a way to guarantee that the task will keep on running. Also the Automatic Update Task can perform load balancing according to the usage rate of processors.

2.3.2 Configuration Script

The configuration script is used in order to configure how the automatic update task will execute the heuristic.

This script defines four variables:

- INITFRAMES: used in order to set how many frames are used by the heuristic to do the initial tests. These initial tests are used so the user may want the heuristic to make the initial tests differently from the normal test;
- DISCARDFRAME: used in order to discard the first DISCARDFRAME frame results, because the main thread can be loading images or models and this can affect the tests;
- LOOPFRAMES: it is used to setup how frequently the heuristic will be executed. If this value is set to -1 the heuristic will be executed only once;
- USEHARDWARE: a variable to determinate which modes will be used for the automatic update tasks;

- EXECUTEFRAMES: it is used to set how many frames are needed before the decision on changing the processor will execute the next tasks.

An example of the configuration script file can be seen in List 1.

List 1: Configuration Variables

```
INITFRAMES ⇐ 20
DISCARDFRAME ⇐ 5
LOOPFRAMES ⇐ 50
USEHARDWARE ⇐ ALLAVAILABLE
EXECUTEFRAMES ⇐ 5
```

The automatic update task begins executing after the DISCARDFRAME are executed. In the sequel, it executes INITFRAMES frames in the CPU cores and the next INITFRAMES in the GPU.

Afterward, it decides where the next LOOPFRAMES frames will be executed. If the LOOPFRAMES is greater than -1, it executes EXECUTEFRAMES frames in the CPU cores and it executes EXECUTEFRAMES frames in the GPU. Finally, it decides where the next LOOPFRAMES frames will be executed and keep repeating until the application is aborted.

The heuristic script is used in order to distribute automatically the tasks between the CPU cores and the GPU. This script defines three functions:

- reset (): reset all the variables that the script uses in order to decide which processor will execute the task. This function is called after the LOOPFRAMES frames are executed. The variable that are normally used by the heuristic are:
  - CPUTime: an array that contains the sum of all the elapsed times that the task has been processed in this CPU thread;
  - GPUTime: the sum of all the elapsed times that the task has been processed in the GPU;
  - numBodies: the number of bodies that have been processed;
  - initialBodies: the number of bodies in the beginning of the processing.
- SetVariable (elapsedTime, numberBodies, processor, thread): this function sets all the variables that the heuristic uses. This function is called after running the EXECUTEFRAMES frames in each processor. The script that defines this function can be seen on List 2.
- main (): This is the function that executes the heuristic and decides which processor will execute the task. This function is called just before the LOOPFRAMES frames is executed.

List 2: SETVARIABLE script

```
numBodies ⇐ numberBodies
if processor == CPU then
CPUTime[thread] ⇐ CPUTime[thread] + elapsedTime
else
GPUTime ⇐ GPUTime + elapsedTime
end if
```

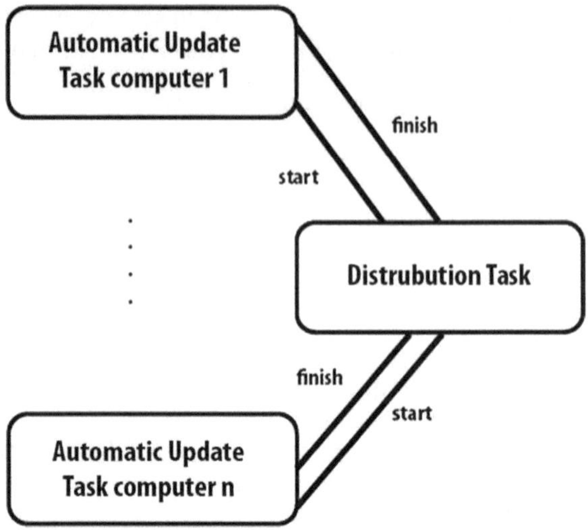

Fig. 2.14 The distribution update task class and messages

The component in the architecture enables the implementation of any real-time loop model and heuristic presented with the adaptation for distribution of tasks between cores of the same processor.

2.3.3 The Distribution Task

The purpose of this class is to define which computer will run the task. The class may change the task's computer during the application execution, which characterizes a dynamic distribution.

One of the major features of this new architecture is to allow dynamic task allocation between the computers. In order to do that it uses the Distribution Task class. This task can be configured in order to be executed in two modes: in the automatic distribution between computer with the information of the network and hardware detected by the Network Check and Hardware Check class, and in a manual mode.

The Distribution Task acts like a server and distribute the tasks between the clients. The role of the automatic update task is to execute a configuration script to determinate the execution mode of it. With that it executes the Network Check, in order to check the network between the computers, and the Hardware Check to know the configuration of each computer. The Distribution task determines which client will execute the next task and will send the task to the chosen client, allowing it to execute. Also, every time the clients finish a task they send the result to the server to let it know it has finished. Figure 2.14 illustrates this process.

Each computer have an Automatic Update Task that receives task from the Distribution Task and executes the task, when it finishes the task execution it sends the results back to the server. If a computer loses the link with the server, the server will automatic distribute its task to another client.

2.3.4 The Configuration Script

The configuration script is used in order to configure how the automatic distribution task will execute the heuristic. This script can be used to define:

- the maximum numbers of computers that will be used in the simulation;
- the computers IP addresses;
- the network mode (if it will run the tasks locally or in the network distribution mode);
- the tasks that can use the network mode;
- the minimum network connection speed between the server and the client in order to use that client machine.

2.3.5 A Summary of the Architecture Execution

First, the server computer (which displays visualization and gathers the player input) queries the network for available computers and their capacity. After that, it divides the amount of work between the computers, considering the processing power of each one. While running the game loop, the server verifies if any computer is down or if there are lost connections, in this case, it tries to redistribute the task to the other remaining computers. It also exchange messages with the computers, process its own tasks that it might have been assigned to, and then presents the results to the end user.

2.4 Conclusion

With the evolution of computer networks, distributing computation will become more in evidence, allowing more complex real-time simulations.

This work presents and summarizes the concept of the main loops present in simulations, a subject that is not very much discussed in the literature. We discuss an architecture for loops that are able to distribute tasks between computers in a network, and inside computers through CPUs, CPU cores, and GPUs. With this approach a simulator is able to use more resources available to it (local and remote), reducing its system requirements.

The framework and concept presented here can be applied to any game or real-time simulation task that can be put in a parallel mode. With the use of a distribution across the internet one could run more processing consuming simulators.

As network bandwidth increases and with the development of cloud computing, the concept of this work presents will become more and more relevant.

Acknowledgments The authors would like to acknowledge CNPq and FAPERJ for the financial support.

References

1. Glinka, F., Ploss, A., Gorlatch, S., Müller-Iden, J.: High-level development of multiserver online games. Int. J. Comput. Games Technol. **5**, 1–16 (2008)
2. Joselli, M., Zamith, M., Clua, E.W.G., Pagliosa, P., Conci, A., Montenegro, A., Valente, L.: An adaptative game loop architecture with automatic distribution of tasks between cpu and gpu. In: Proceedings of the VII Brazilian Symposium on Computer Games and Digital Entertainment, pp. 115–120 (2008)
3. Joselli, M., Zamith, M., Valente, L., Clua, E.W.G., Montenegro, A., Conci, A., Feijó, B., Dornellas, M., Leal, R., Pozzer, C.: Automatic dynamic task distribution between CPU and GPU for real-time systems. In: IEEE Proceedings of the 11th International Conference on Computational Science and Engineering, pp. 48–55 (2008)
4. Joselli, M., Clua, E., Montenegro, A., Conci, A., Pagliosa, P.: A new physics engine with automatic process distribution between CPU–GPU. In: Sandbox 08: Proceedings of the 2008 ACM SIGGRAPH Symposium on Video Games, pp. 149–56 (2008)
5. Joselli, M., Zamith, M., Clua, E., Leal-Toledo, R., Montenegro A., Valente, L., Feijó, B., Pagliosa, P.: An architecture with automatic load balancing for real-time simulation and visualization systems. J. Comput. Interdiscip. Sci. **5**, 207–224 (2010)
6. Joselli, M., Zamith, M., Clua, E., Pagliosa, P., Conci, A., Montenegro, A., Valente, L.: (2008) An architecture with automatic load balancing and distribution for digital games. In: Proceedings of the IX Brazilian Symposium on Computer Games and Digital Entertainment, pp. 59–70
7. Valente, L., Conci, A., Feijó, B.: Real time game loop models for single-player computer games. In: Proceedings of the IV Brazilian Symposium on Computer Games and Digital Entertainment, pp. 89–99 (2005)
8. Dalmau, D.S.C.: (2003). Core Techniques and Algorithms in Game Programming. New Riders Publishing, US
9. Dickinson, P.: Instant replay: Building a game engine with reproducible behavior. http://www.gamasutra.com/view/feature/3057/instant_replay_building_a_game_php (2001). Accessed 2 March 2014
10. Watte, J.: Canonical game loop. http://www.mindcontrol.org/~hplus/graphics/game_loop.html (2005). Accessed 2 March 2014
11. Gabb, H., Lake, A.: Threading 3D game engine basics. http://www.gamasutra.com/view/feature/2463/threading_3d_game_engine_basics.php (2005). Accessed 2 March 2014
12. Monkkonen, V.: Multithreaded game engine architectures. http://www.gamasutra.com/view/feature/130247/multithreaded_game_engine_.php?print=1 (2006). Accessed 2 March 2014
13. Valente, L, Conci, A., Feijó, B. Real time game loop models for single-player computer games. In: Proceedings of the IV Brazilian Symposium on Computer Games and Digital Entertainment, pp. 89–99 (2005)

14. Microsoft: Documentation. http://msdn.microsoft.com/en-us/library/bb200104.aspx (2006). Accessed 2 March 2014
15. Rhalibi, A.E., Costa, S., England, D.: (2005) Game engineering for a multiprocessor architecture. In: DIGRA Conference
16. Kerr, A.: Campbell, D., Richards, M.: (2008) GPU VSIPL: High-performance VSIPL implementation for GPUS. High Performance Embedded Computing (2008)
17. Bolz, J., Farmer, I., Grispun, E., Schrder, P.: Sparse matrix solvers on the GPU: conjugate gradients and multi-grid. ACM Trans. Graph. **22**(3), 917–924 (2003)
18. Ufimtsev, I.S., Martnez, T.J.: Quantum chemistry on graphical processing units. 1. Strategies for two-electron integral evaluation. J. Chem. Theory Comput. **4**(2), 222–231 (2008)
19. Nyland, L., Harris, M., Prins, J.: Fast n-body simulation with cuda. GPU Gems 3 Chapter 31, pp. 677–695 (2007)
20. Passos, E., Joselli, M., Zamith, M., Rocha, J., Montenegro, A., Clua, E., Conci, A., Feijó, B.: Supermassive crowd simulation on GPU based on emergent behavior. In: Proceedings of the VII Brazilian Symposium on Computer Games and Digital Entertainment, pp. 81–86 (2008)
21. Ierusalimschy, R., de Figueiredo, L.H., Celes, W.: Lua 5.1 Reference Manual. Lua.org (2006)

Chapter 3
GIDE: Graphic Interface for Discrete Element

Harold Trannois, Jérome Fortin, Cyril Drocourt and Frédéric Dubois

Abstract In this chapter we propose a graphic display tool for the results of calculations carried out using a discrete element code: Graphic Interface for Discrete Element Code (GIDE). This is a post-processing application written in C++ based on portable open source libraries, making GIDE compatible with different OS (Windows, Linux, Unix, MacOS, etc.).

Keywords File format · Distinct element method · Visualisation · Post-processing

Abbreviations

API	Application programming interface
DEM	Distinct element method
FEM	Finite element method
FLTK	Fast light toolkit
GLUT	OpenGL utility toolkit
HDF	Hierarchical data format

H. Trannois · J. Fortin · C. Drocourt (✉)
Laboratoire des Technologies Innovantes EA 3899, INSSET, Université de Picardie Jules Verne, 48 rue Raspail, 02100 ST Quentin, France
e-mail: cyril.drocourt@u-picardie.fr

H. Trannois
e-mail: harold.trannois@u-picardie.fr

J. Fortin
e-mail: jerome.fortin@u-picardie.fr

F. Dubois
CNRS/Université Montpellier 2, Laboratoire De Mécanique et Génie Civil (LMGC), Montpellier, France
e-mail: dubois@lmgc.univ-montp2.fr

F. R. Leta (ed.), *Visual Computing*, Augmented Vision and Reality 4,
DOI: 10.1007/978-3-642-55131-4_3,

HDF5 Hierarchical data format (version 5)
OSG Open scene graph
XML Extensible markup language

3.1 Introduction

Traditionally, when carrying out a numeric investigation of the mechanical behaviour of a deformable body undergoing several stresses, use is made of the Finite Element Method (FEM), which is an important tool in the analysis of structures and more generally in engineering science [1]. It is based on the mechanics of continuous media; however, the bodies are not continuous, but the assumption of continuity affords a simplification making it possible to solve the problems of classical mechanics. However, the assumption of continuity appears difficult to accept for systems composed of several rigid or deformable parts, interconnected by links. We then refer to multi-body systems. Currently, numerous applications involve the study of such systems. In the field of sport, we study the movements of athletes. In civil engineering, the modelling of granular materials by a multi-body system enables understanding of the origin of mechanical behaviour, whether it is microscopic, macroscopic, etc. In the field of the automobile and transport, we seek continually to improve the performance, comfort, and safety of cars, lorries, and trains. In granular mechanics, geomaterials or masonry, numerical simulations based on the individual behaviour of grains or blocks are qualified as DEM or Distinct Element Method [2], in contrast to the FEM strategy used when a homogenised behaviour law has been chosen, assimilating the granulate or masonry to a continuous medium.

For the moment in granulate mechanics, because of limitations in memory size and calculation time, discrete numerical simulations are limited to samples of a few thousand, or even a few tens of thousands of grains. For comparison, 1 cm^3 of sand with 0.1 mm diameter grains contains about 10^6 grains. Thus, another substantial problem in modelling granular media is to be able to define the average magnitudes (average stresses, average strains, etc.), taking into account the overall behaviour of a granular medium considered as a continuous medium, and representative of the physics at the scale of the grain (contact efforts, volume efforts, local rotations and speeds, etc.). The final aim is to obtain homogenised behaviour laws. This is the object of the micro-mechanical approaches that can be enriched with the results obtained by discrete numerical simulations of granular media, some quantities such as intergranular forces being difficult to measure experimentally.

A modelling problem is generally characterised by defining a real physical system in which certain quantities are a priori unknown, and others assumed to be known. The first step in the modelling process consists in making a series of simplifying assumptions that make it possible to model the problem: idealisation of the geometry, boundary conditions and stresses. The second step consists in selecting the relations that govern the model (taking account of friction, shock law, thermal effect, wearing phenomena; or remote interactions), eliminating certain variables between these relations, making simplifying assumptions (rigidity of bodies), then choosing the methods of discretising the equations thus obtained. Creating such a discretised model uses numerous implicit or explicit knowledge of the user: choice of time step, choice of stiffness, choice of coefficients of restitution, etc. We can thus obtain a system of algebraic equations, which approximately represents the behaviour of the physical system being investigated [3].

Solving the system supplies the unknowns, here the speeds and local reactions. The development of a numerical tool for displaying in 2D and 3D of mechanical systems in unilateral dynamics, of large size, i.e. containing more than 10,000 particles, naturally involves a series of conditions to be satisfied by the model:

- Portability: it must be easy to integrate and to use in divers calculation environments;
- Performance: it must have minimum cost while enabling faithful modelling of the mechanical part of the system;
- Reliability: it must be accurate and robust so as to represent as accurately as possible the behaviour of the mechanical system whatever the conditions to which it is submitted and the time interval that is simulated.

The aim of this chapter is to propose a graphic display tool for the results of calculations performed with the aid of a Discrete Element code [4]: Graphic Interfaces for Discrete Elements (GIDE). This is a post-processing application based on portable open source libraries, making GIDE compatible with various OS (Windows, Linux, Unix, MacOS, etc.). GIDE is a vector tool; this alternative approach to current display tools allows the discrete aspect of bodies to be conserved; two bodies interconnected by a third are not transformed into a 3D image, they remain 3 graphically representative elements that can be selected individually.

The choice of libraries has been a determining factor, whether with HDF [5] for the handling of data files or OpenSceneGraph (OSG) [6, 7] for managing of 3D scenes in OpenGL.

GIDE is also a post-processing application enabling a body to be tracked in time or data to be extracted with the aid of the tool called 'capteur' in reference to the tool that can be found when experiments are carried out.

Finally, GIDE is equipped with a filter allowing the importing/conversion of data files from various calculation codes (MULTICOR [4], LMGC90 [8]).

3.2 Technology Choice

GIDE was designed to be as open as possible through recourse to recognised and free libraries. We have been particularly attentive to the documentation of the code and the development environment through collaborative work. The tools used are all from open sources. The result is an application entirely uncoupled from EE calculation code.

3.2.1 *Data Format*

3.2.1.1 First Version of the Data Format

In order not to be intrusive, the initial aim of GIDE was to be capable of functioning without any modification of upstream software, such as MULTICOR or other simulators, only via the transformation of data from these applications. The transformation is based on an XML file describing the organisation of the data and a conversion tool (included in the GIDE). Thus, all results data files are transformed into the format HDF5 before use.

The format HDF5 was adopted as the native GIDE format; its advantages are:

- A data model allowing the representation of complex data,
- A portable file format,
- A library that can be used on different platforms, with interfaces in C, C++, Fortran and Java,
- Optimised performance for access time and size of data,
- Tools for the display and processing of data in the format HDF.

Reading and writing are optimised in order to exploit various types of architecture: simple file on a standard file system, several files on a standard file system, several files on a parallel file system and other situations. The first version of the organisation of data in the HDF file offers the following architecture (Fig. 3.1).

Starting from the root '/' w access the simulation of global data, and then we access each scene and finally the block of data for each type of discrete element. In the HDF file there are sets of data called DATASET, which are gathered into groups called GROUP. Each GROUP or DATASET has a name; it is the name of the DATASET that determines the representation of a discrete element. The tree representing the organisation of the data looks like Fig. 3.1. Access to the first DATASET is via path/scene/pas01/. The data are not only organised in the form of a tree in the files, but also in the memory. OSG also organises the various elements of a 3D scene in the form of a tree and even more in the form of a graph. GIDE re-uses the mechanisms of OSG for this part; the discrete elements of GIDE are specialisations of the GROUP of OSG. While data are being read, the tree is built up of the scene for each step in the simulation, Fig. 3.2.

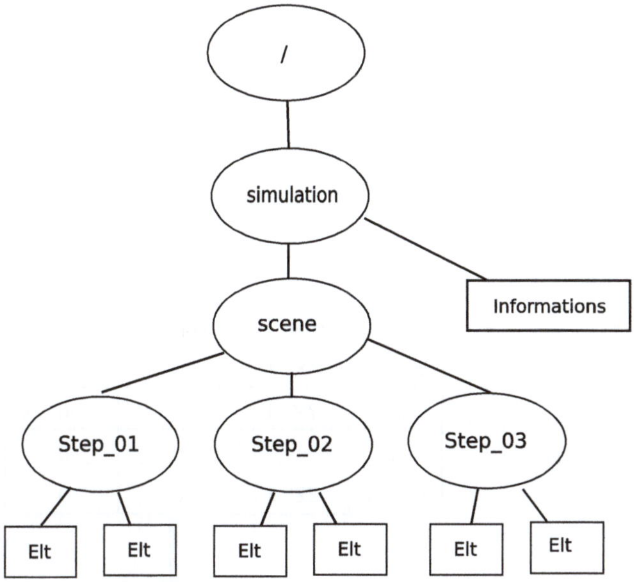

Fig. 3.1 HDF representation

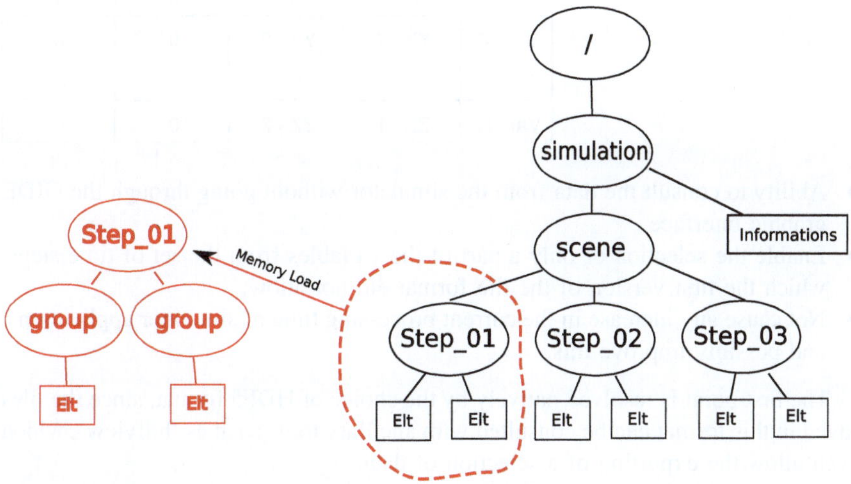

Fig. 3.2 Representation in the memory

3.2.1.2 Second Version of the Data

The second phase in the development of the GIDE project was to integrate the recording of data directly in the format HDF5 into the simulator application such as MULTICORPS. The constraints were as follows:

Fig. 3.3 Representation matrix

	Step 1	Step 2	...	Step M
Var 1				
Var 2				
...				
Vat N				

Fig. 3.4 Internal representation 1

	Step 1	Step 2	...	Step M
Var 1	XX 1	0	0	0
Var 2	YY 1	0	0	0
...	...	0	0	0
Var N	ZZ 1	0	0	0

↓

	Step 1	Step 2	...	Step M
Var 1	XX 1	XX 2	0	0
Var 2	YY 1	YY 2	0	0
...	...	...	...	...
Var N	ZZ 1	ZZ 2	0	0

- Ability to consult the data from the simulator without going through the GIDE graphic interface,
- Enable the selection of only a part of the variables from the set of time steps, which the first version of the file format did not allow,
- Not cause any increase in the current processing time of simulator applications, and possibly improve this.

The first point is resolved natively by the choice of HDF5 format, since the files saved in this format can be consulted with ancillary tools such as 'hdfview', which even allow the exporting of a selection of data.

The second point made us think about an alternative in the representation of data in the HDF5 file, and the solution adopted is a matrix Fig. 3.3.

With the possibilities afforded by the DATASET of the HDF files, there are then two solutions possible:

- Build the matrix initially with the desired size, from the start of the programme, and update it at each time step (Fig. 3.4),
- Use a resizeable matrix of initially empty size, and expand it at each time step, with the size of a vector of dimension [1: N] (Fig. 3.5).

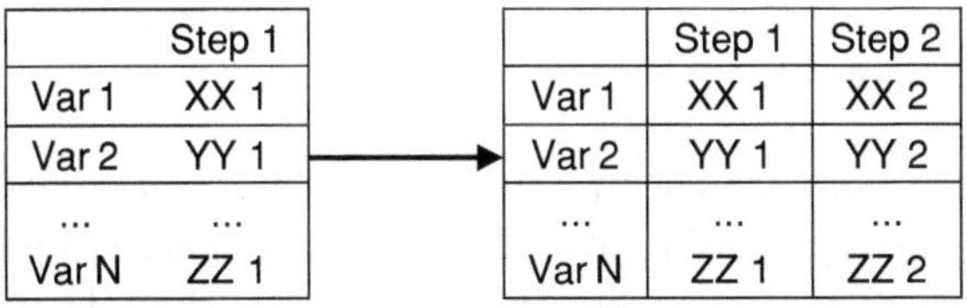

Fig. 3.5 Internal representation 2

Table 3.1 Results of the tests

	A	B	C	D	E
Real	0 min 4.596 s	0 min 0.351 s	0 min 0.974 s	11 min 36.040 s	0 min 0.557 s
User	0 min 4.023 s	0 min 0.001 s	0 min 0.485 s	0 min 9.820 s	0 min 0.227 s
Sys	0 min 0.417 s	0 min 0.292 s	0 min 0.411 s	7 min 1.452 s	0 min 0.292 s

For a solution to the third point a series of tests on the various solutions had to be devised, and the best solution selected. The possible solutions are as follows:

- Write the data in text form, using standard recording primitive (current solution);
- Write the data in raw binary form, using the standard recording primitives (reference solution);
- Use the HDF5 format of the first version of the GIDE;
- Use an initialised matrix at maximum size;
- Use a matrix of empty size, increasing it by one vector at each time step.

Each of these solutions was tested with the recording of 4,096 variables of the long integer type (32 bits), over 4,096 time units. These tests were carried out with the 'time' command under Linux, and the result is given in the Table 3.1.

The good results from solution B are due to the possibility in binary mode of recording a vector in a single operation, stating the memory address and the length of this vector. However, this solution does not satisfy the prerequisites overall.

Aside from this point, it clearly appears that solution 'E' gives the best results, and this is therefore the solution selected and integrated in the GIDE. Moreover, it offers the advantage of satisfying our three initial constraints.

3.2.2 3D Motor

The choice of 3D library is crucial. Though this interface is considered as a navigation tool in an ocean of data, the first contact with the user is nevertheless visual and this is his first expectation: to display the results of the simulation. There was no question of making an nth 3D library, but rather of choosing the most appropriate one from existing ones. Four constraints were set:

3.2.2.1 First Constraint: Targeting the Material and Operating System to the Users

Most workstations are individual laptops; these are not supercomputers or CAD stations and the post-processing of results is for the moment a poorly furnished second order activity. The use of the API 3D standards ‘Direc3D’ (by Microsoft ®) or ‘OpenGL’ (by Khronos Group) is therefore strongly indicated! They allow currently available graphic cards to be driven and use to be made of their internal calculator thereby leaving the main processor free.

For the operating system, we decided on portability over the three commonly used OS: ‘Linux ®’, ‘Windows ®’, ‘MacOS ®’. This constraint eliminates the API ‘OpenGL’, which is for the moment only supported by ‘Windows’. GIDE therefore uses the API ‘OpenGL’ to drive the graphic cards.

3.2.2.2 Second Constraint: Displaying a Large Number of Elements

This constraint turns out to be difficult to respect. It is in contradiction with the first. Without making use of a dedicated display machine a large number of bodies cannot be displayed. The threshold of 10,000 is often the limit beyond which manipulation of the display becomes jerky and even blocks for small configurations. It is in the organisation of the bodies that we hope to overcome this constraint. To do this we use graphs of scenes, techniques developed initially for an IRIS project [9] and then taken up by one of its authors in an OSG project involving more general use [10].

The data from the discrete element calculation code are particularly well suited to the construction of graphs of scenes since the latter are made up of vector elements: circles, rectangles, spheres, cylinders, polyhedral, etc., and it is these same basic elements that are used in simulations of the DEM type.

One of the major strengths of this approach lies in the elimination of calculations for masked bodies and other elements of the scene, called culling [11], and it is this strength that is particularly interesting in our case. From experience, DEM simulations have bodies organised in heaps, which means that most of the bodies need not be represented as those on the surface mask them. For example, if we take a cube of 50 bodies alongside, the simulation will count $50^3 = 125{,}000$ bodies for only three visible faces on a maximum of two rows, i.e. $50 \times 50 \times 2 = 15{,}000$ bodies, which is the limit of manipulability on a laptop (Fig. 3.6).

3.2.2.3 Third Constraint: Select the Bodies

The currently available offer of applications for displaying FEM is very rich, unlike that for DEM, which is very poor.

What is the difference between the two, which justifies such a disparity?

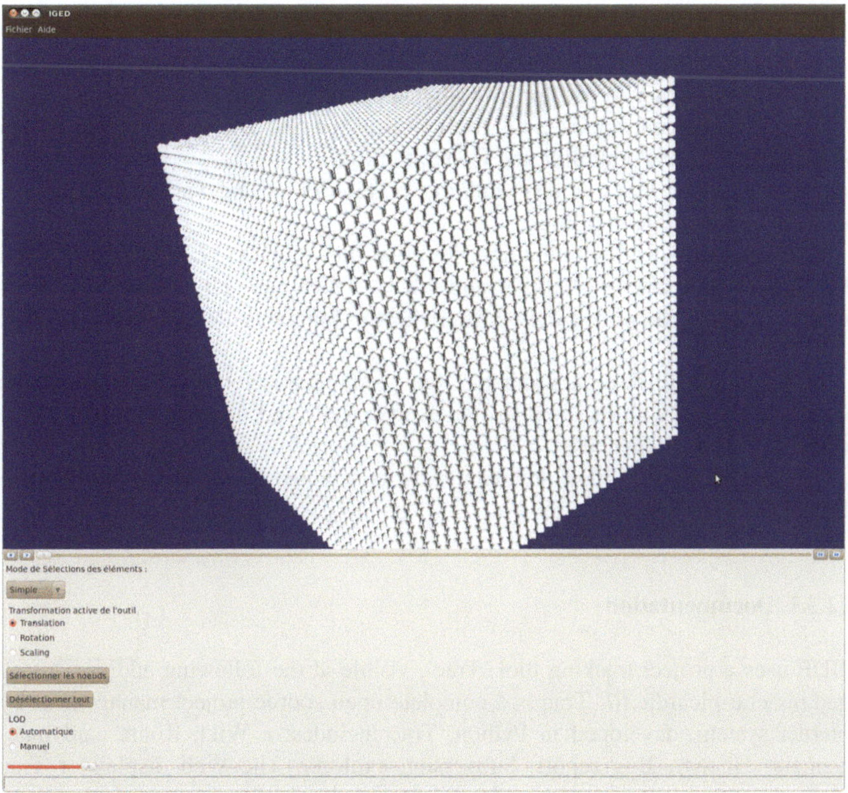

Fig. 3.6 Cube in 3D (40,000 bodies)

There is first of all the novelty of DEM, which has only been used for a few years, but also the unsuitability of methods for representing discrete finite elements. It is for example impossible to select several 3D bodies graphically since the notion of group of nodes forming a body is in contradiction with the spirit of meshes used in FEM.

Representation in the form of vector elements is also a good response to this constraint! This reinforces the choice of a data structure in the form of a graph as in 'AutoCAD ®', 'Adobe Illustrator ®', 'Acrobat 3D ®', etc.

In conclusion to this section, we stress the fact that GIDE is closer to to a vector image application than bitmap streaming. This explains why we will not further refer to remote display solutions [11] suitable for representing a very number of data, which only partially meet our requirements.

3.2.3 *The User Interface/Documentation*

The choice of library to use for user interface management: menu, dialogue box, window management, mouse, etc. must respect the above constraints, which can be summarised as: Portability and Compatibility.

The choice of OSG library and the portability of the various OS (cross-platform) limit the possibilities.

OSG interfaces well with the libraries or toolkits: GLUT, FLTK, QT, GTK and WxWidgets and also with FOX and MCF, which are not themselves cross-platform. Among these five only those richest in functions are selected: QT, GTK and WxWidgets.

At this stage all three are suitable, but preference is for WxWidgets with its better integration with the host 'windows manager', and the look and feel of the native system are retained

To summarise, the GIDE architecture is represented by the following diagram (Fig. 3.7).

3.2.3.1 Documentation

GIDE uses a project tracking tool 'Trac', visible at the following address: http://iged.insset.u-picardie.fr/. Trac is a complete open source project management by internet system, developed in Python. Trac includes: a Wiki, Route card management, History, Bug report, Subversion explorer. The Web display of Trac works through an engine in the ClearSilver template.

3.3 Characteristics of the Application

In the previous chapter, we saw that GIDE is a vector type application allowing the selection of simulation elements as unitary entities. This characteristic is present in design applications such as computer-aided design or drawing.

To make things clear, take two flagship applications from Adobe: Illustrator and Photoshop. The former, vectorial, is used to create illustrations, the patter, raster, to retouch images. Vector scan is used for creation and raster scan for display.

The scientific data visualisation application 'ParaView' (Open-Source software) produces 3D without notions of vectors. So why do we proceed as we do? Why process the image as a set of vectors when there is no question of modifying the simulation—we have no right to change the data!

The idea is actually quite different here; the aim is not to allow modification of the image, but to allow the bodies to be tracked in time, by tracing them or, for example, by numbering them within a particular zone.

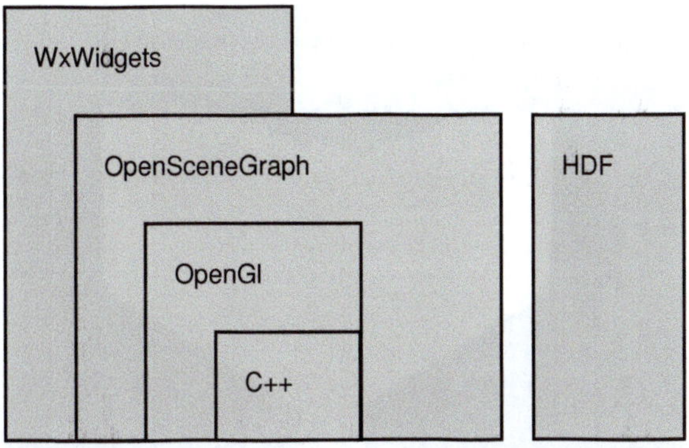

Fig. 3.7 GIDE architecture

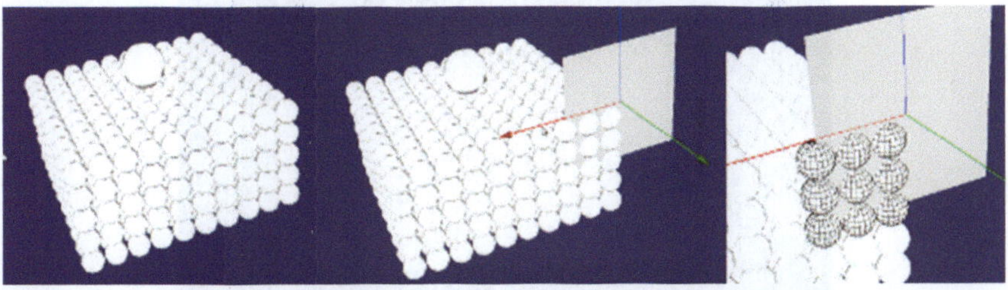

Fig. 3.8 Showing marking of an element

3.3.1 Tracer

In GIDE, a tracer is a visual marker applicable to a body. It enables one or more bodies to be followed throughout the simulation. To apply a marker a zone of space is selected. All bodies within this zone will be marked (Fig. 3.8).

The user has several tools at his disposal, to delimit a zone of space: {the point, the line, the plane, the cubic volume}. Each of these tools is directly manipulable with the aid of the mouse. Marking is then done by intersection of the zone with the set of bodies; all the bodies having an intersection, even partial, with the defined zone will be marked.

The following example illustrates the marking of a layer in a silo, and its follow-up during the flowing (Figs. 3.9, 3.10, 3.11 and 3.12).

3.3.2 Sensor

A sensor is defined as an active element. With it, data can be extracted and calculations performed on them. This part is currently limited to applying a calculation formula to each of the bodies encountered by the sensor and to trace the result.

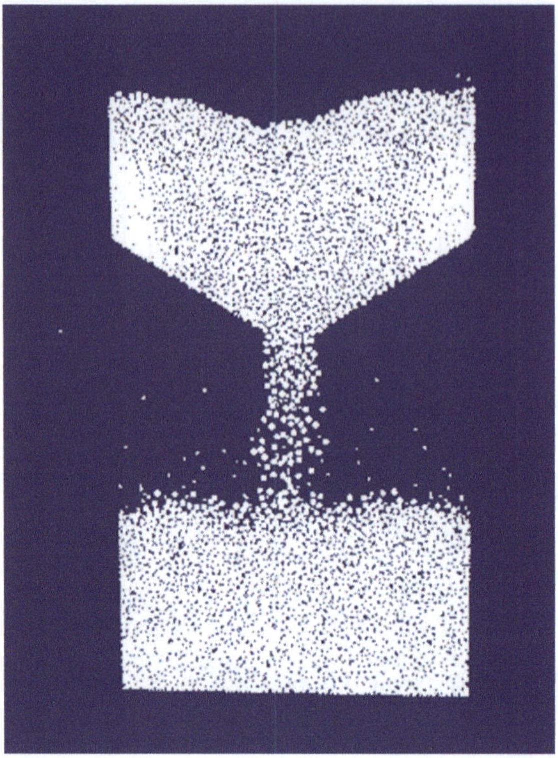

Fig. 3.9 Emptying of a silo

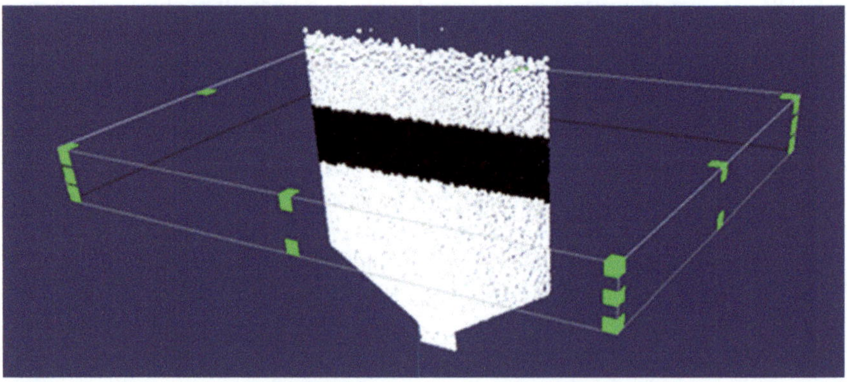

Fig. 3.10 Marking bodies by intersection of volume

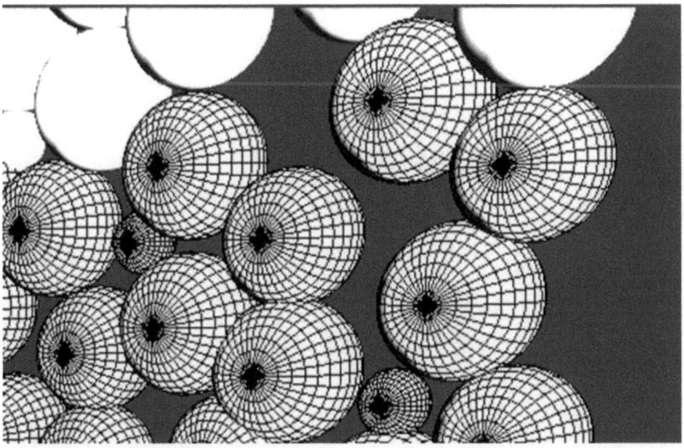

Fig. 3.11 Detail of marking of bodies

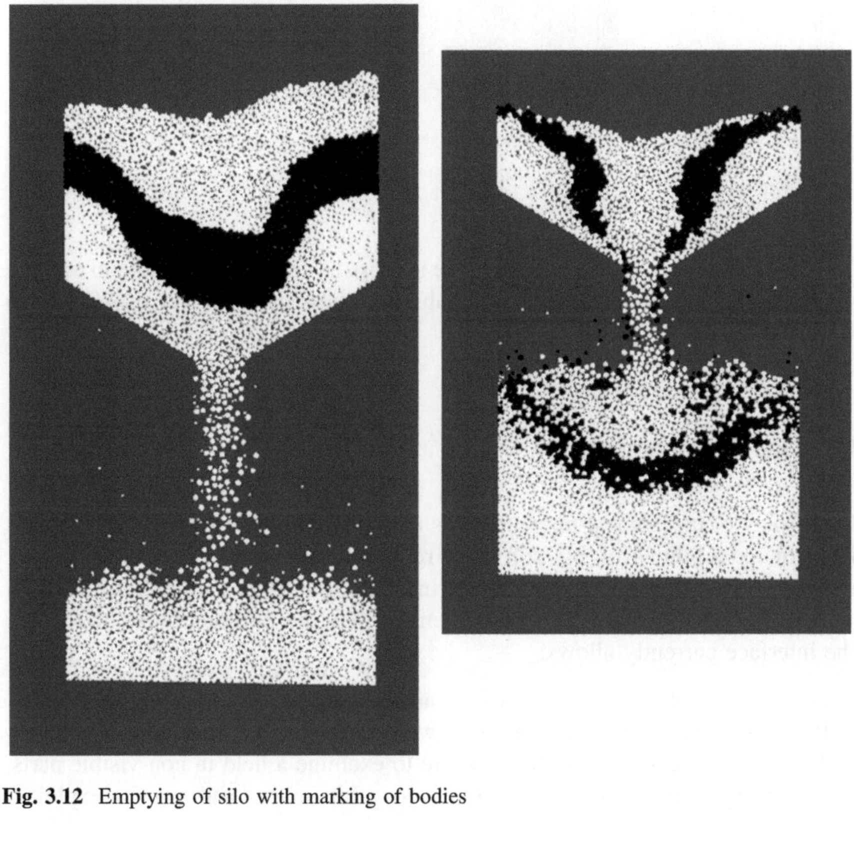

Fig. 3.12 Emptying of silo with marking of bodies

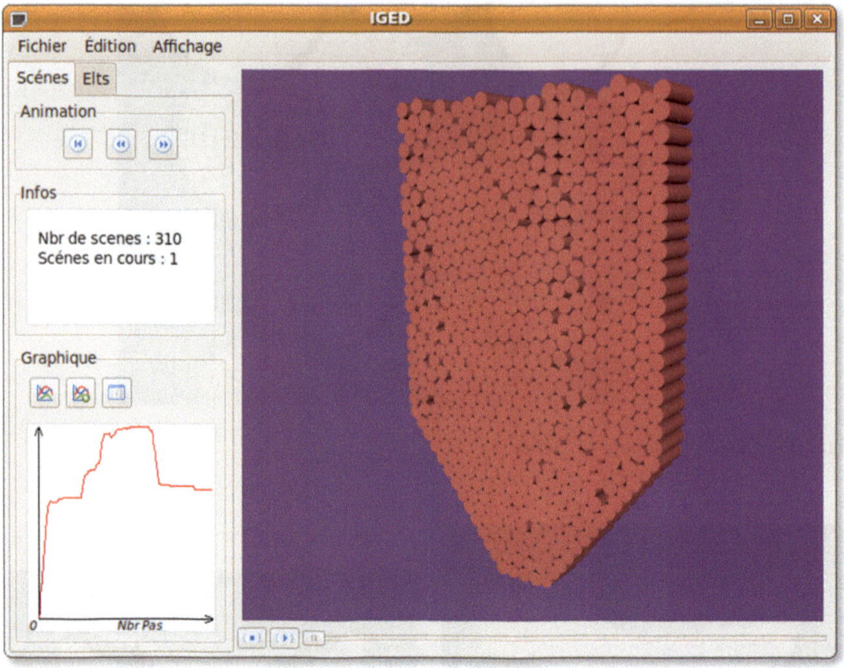

Fig. 3.13 Curve

Developments are in hand to script the sensors. A sensor will be able to execute a script (python), and the latter will be able to access the body via an API and to produce outputs (Fig. 3.13).

3.4 The Representation

The display of mechanical phenomena is the most important part. GIDE includes the main possibilities of 3D software: rotations, displacements, etc. We have, however, added the capability of following a body in time and tracing the associated curves: we can thus at any moment display the information on this body. The interface currently allows:

- 3D display—of isovalues—in wire frame/hidden face mode,
- Rotation, zoom, dynamic translation with the aid of the mouse,
- The ability to cut through the structure to examine a field in non-visible parts,
- Recording of mpeg formats for films, jpeg for images and svg for exploited data.

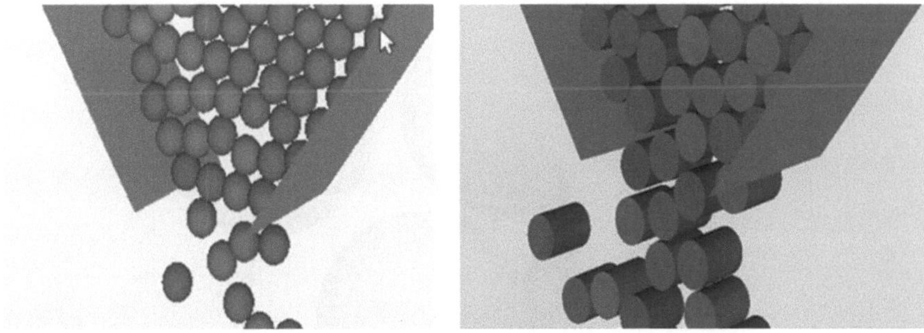

Fig. 3.14 Example of change of representation

3.4.1 Flexibility of Representation

In GIDE, a discrete element is an indivisible entity, and it is a unitary element. It is thus possible to select it, to manipulate it, to track it, to decorate it, etc. This approach is very different from representations by meshing. It more closely matches the granular world and enables the virtual experimental medium to be observed and dissected, and its characteristics extracted for comparison with theory.

The drawing of an element in GIDE is done via plugins, affording a representation library adapted to each phenomenon under study. By default the name DATASET is the name of the plugins giving, for example, the possibility of changing the representation during display. Some simulations are done in 2D or the body is considered mechanically as a cylinder when it is actually comprised of spherical elements; in this case, the user needs to change the representation according to the desired approach.

The change in representation is done according to the available plugins. The following figures illustrate the example of the spherical representation becoming cylindrical (Fig. 3.14).

3.4.2 Representation of Physical Data

Apart from displaying the position of a body, a researcher often studies other physical phenomena: temperature, speed, electric potential, etc. He should therefore be able to display these data. The notion of decorators has been implemented in GIDE; it is based on Design Pattern. It allows the graph representing the scene to be modified by adding nodes. The modifications are cumulable with each other. The representation of forces can, for example, be activated with that of temperatures (Fig. 3.15).

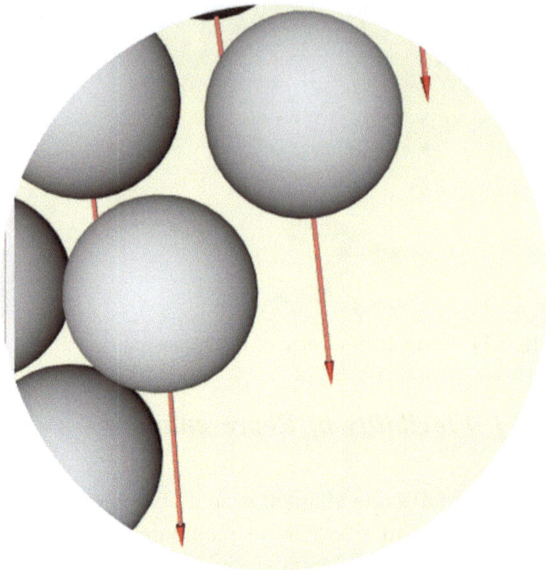

Fig. 3.15 Speed vector

3.5 Conclusion

The language C++ was chosen for the development. With this language the OpenGL graphic card capacities can be fully exploited, which is necessary for the processing of 3D scenes rich in bodies. Moreover, as libraries such as OSG are also written in C++, we have been able to exploit them and use them to the full via the inheritance. The tree data structure of the HDF file has allowed the implementation of a cache; the cache is managed using threads that allow configurable number, until now have smooth animations.

GIDE has a thread safe cache and OSG is also thread safe. The graphic interface has been developed with the aid of the WxWidget toolkit portable on all the OS. This is a display tool for better exploitation of data from a Discrete Element code. It provides a representation that supports the researcher's discourse. Finally, respecting the design patterns during development should allow other developers a relatively rapid learning curve. The presentation of GIDE will be through various digital applications from the code ED MULTICOR developed in the Laboratoire des Technologies Innovantes and the code LMGC90 developed in the LMGC of Montpellier.

References

1. Zienkiewicz, O.C.: The finite element method, 3rd edn. McGraw-Hill, New York (1977)
2. Cambou, B., Jean, M.: Micromécanique des matériaux granulaires. Hermès Science, Paris (2001)
3. Duran, J.: Sables, poudres et grains. Introduction à la physique des milieux granulaires, Eyrolles sciences (1997)
4. Fortin, J., Millet, O., de Saxcé, G.: Numerical simulation of granular materials by an improved discrete element method. Int. J. Numer. Meth. Eng **62**, 639–663 (2005)
5. Hierarchical Data Format (HDF). http://www.hdfgroup.org
6. 3D Graphics Toolkit. http://www.openscenegraph.org/projects/osg
7. OSG. http://www.openscenegraph.org
8. Dubois, F., Jean, M.: Lmgc90 une plateforme de développement dédiée à la modèlisation de problèmes d'intéraction. 6 ème colloque national en calcul des structures 111–118 (2003)
9. Rohlf, J., Helman, J.: IRIS performer: a high performance multiprocessing toolkit for real-time 3D graphics. SIGGRAPH '94: Proceedings of the 21st Annual Conference on Computer Graphics and Interactive Techniques (1994)
10. Staneker, D., Bartzb, D., Straßera, W.: Occlusion Cullingnext term in OpenSG PLUS. Comput. Graph. **28**, 87–92 (2004)
11. Limet, S., Madougou, S., Melin, E., Robert, S.: Rapport de Recherche. http://www.univ-orleans.fr/lifo. La visualisation distante, Université d'Orléans LIFO, Rapport N° 2006-12, 20/12/2006 (2006)

Chapter 4
Important Parameters for Image Color Analysis: An Overview

Juliana F. S. Gomes, Fabiana R. Leta, Pedro B. Costa and Felipe de O. Baldner

Abstract In recent years it is noteworthy how the use of Computational Vision techniques in processing and quality control of products has advanced. The available resources in both electronic and computing were important factors in the automation development, allowing constant monitoring during the process. Such techniques have systematically evolved in the international commerce. However, there is a lack of standardization on quality control of products using image analysis. Measurements using digital image should consider important aspects, such as the effects of lighting, characteristics of the environment, the types of illuminants, the observers, to name a few, all that beyond the traceability of the system and the definition of standards. With this in mind, the aim of this chapter is to discuss the relevance of the main variables that influence the color measurement of images using computer vision techniques, in order to promote some thought about the needs of standardization.

Keywords Color analysis by image · Illumination · Color perception · Color rendering index · Color temperature

J. F. S. Gomes (✉) · P. B. Costa · F. de. O. Baldner
Instituto Nacional de Metrologia, Qualidade e Tecnologia—Inmetro,
Duque de Caxias, RJ 25250-020, Brazil
e-mail: jfgomes@inmetro.gov.br

P. B. Costa
e-mail: pbcosta@inmetro.gov.br

F. de. O. Baldner
e-mail: fobaldner@inmetro.gov.br

F. R. Leta
Dimensional and Computational Metrology Laboratory, Mechanical Engineering Department, Universidade Federal Fluminense—UFF, Niterói, RJ 24210-240, Brazil
e-mail: fabianaleta@id.uff.br

F. R. Leta (ed.), *Visual Computing*, Augmented Vision and Reality 4,
DOI: 10.1007/978-3-642-55131-4_4,

List of Abbreviations

CVS	Computer vision systems
LED	Light-emitting diode
CT	Color temperature
CCT	Correlated color temperature
CCD	Charge coupled device
CMOS	Complementary metal oxide semiconductor
CIE	*Commission Internationale de l'Eclairage* International Commission on Illumination
CRI	Color rendering index
SPD	Spectral power distribution
RGB	Red, green, blue color system
HSL	Hue, saturation, lightness color system
HSI	Hue, saturation, intensity color system
HSB	Hue, saturation, brightness color system
HSV	Hue, saturation, value color system

4.1 Introduction

Visual inspection, despite being a technique that dates from centuries ago, is still a widely used nondestructive test that is able to assess the quality conditions of materials due to its ease of performing, inexpensiveness and generally requirement of no special equipment. This technique is considered a primary method in quality control programs needing an observer with good vision and technical experience in the recognition of defects, as well as good lighting conditions.

With the development of computer software, digital cameras, computers with greater resources and interfaces, together with a decrease in prices, and better access to new technologies, the field of Computer Vision has emerged as a tool for developing innovative methods of nondestructive testing with applications in the most diverse areas. The use of automated inspections in industries is becoming an increasingly attractive solution, especially for compliance analysis, defects search, and final analysis of product quality. Unlike the problems presented for visual inspections performed by humans, these types of systems can provide repeatable measurements without any kind of physical contact, eliminating aspects such as subjectivity, fatigue, and cost associated with human inspection as well as increasing how fast each product is inspected. These comprise many reasons for the use of such new methods of measurement, which may involve not only dimensional aspects, but also appearance characteristics, such as in food, textiles, cosmetics, paints, and automotive, encompassing all industrial sectors.

The use of automated inspections in industries has become an attractive solution in the final analysis of product quality, which considers the characteristics of color, texture, and shape, which are the usual criteria used by consumers when purchasing a product [1, 2]. The main challenge for these inspection systems is to combine the image quality results with lower costs and losses in the process, especially considering the accuracy and reliability of the process. Because of this, each system should be developed and configured properly in order to consider its reproducibility and traceability, making it a challenge for the industry, requiring a greater interface between the professionals of the fields of Engineering, Metrology and Computation.

This work has the purpose of discussing the main parameters that influence color measurement in images using computer vision techniques.

4.2 Computational Vision Systems

Computer Vision is the science responsible for the study and application of methods that enable computers to understand the contents of an image and interpret important features extracted from this image for a particular purpose [3].

The development of Computer Vision Systems (CVS) requires an input data that is usually obtained from sensors, cameras, or videos, which is an image. The image is then processed and transformed into some sort of expected information. Even though a CVS should be organized according to its application, there are typical steps for all CVS, which can be summarized as: image acquisition, pre-processing, segmentation, feature extraction, and processing (analysis); all of these are shown in the block diagram of Fig. 4.1.

In the image acquisition step, the image of a real object or a scene is transformed into a digital image using an acquisition device (digital cameras, scanners, videos, etc.). To represent and manipulate these digital images it is necessary to create mathematical models suitable for this purpose. These are constructed from an image of the real object, which undergoes a transformation in order to be verified. This scanning process generates a continuous mapping of the actual image, which is discretized at various points, called pixels. A matrix is then formed in a way that each position (x, y) that has information on the gray level or color associated with $f(x, y)$. The color is represented by color systems and the most widely used systems are the RGB and HSL.

The preprocessing is the step prior to feature extraction, which aims to improve the acquired image. It can enhance visibility and the separability between the background and the objects, without adding information to the image. Among the techniques of preprocessing, it is possible to highlight the transformation to grayscale, as well as thresholding and filtering.

After the preprocessing there is the segmentation step whose purpose is to divide an image into homogeneous units, considering some of its essential characteristics, for example, the gray level of the pixels and texture contrast. These units are called regions and may correspond to objects in a scene, and are formed by a group of pixels with similar properties. Through this process of dividing an image into regions, which will simplify and/or change its representation, it is

Fig. 4.1 Typical steps of a computer vision system

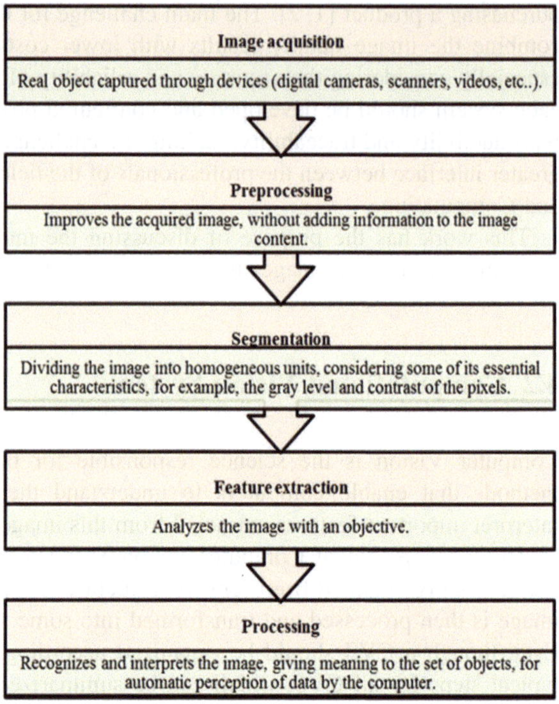

possible to interpret it more easily. After these steps, the image should be ready to have its characteristics of interest obtained.

Lastly, the processing step recognizes patterns and interprets the image, giving meaning to the set of objects in the image, in order to improve on human visualization and the perception of the data automatically by computer.

Currently, computer vision has been used in various industry sectors, contributing significantly in many stages of the production cycle of a product, such as robot control, task automation, production planning, and quality control. In terms of hardware, a CVS is basically composed of a light source, a device for image capture and a computer system, as shown in Fig. 4.2. A CVS is usually used in production lines where human activity is repetitive and tedious. In this situation, products are manufactured very quickly requiring fast and accurate measurements, necessary for making decisions during the process.

4.3 The Color Perception

Currently, color theory can be divided into three main sorts: physical color (physical optics), physiological color (physiological optics), and chemical color (physicochemical optics). Thus, the phenomenon of color depends on the nature of

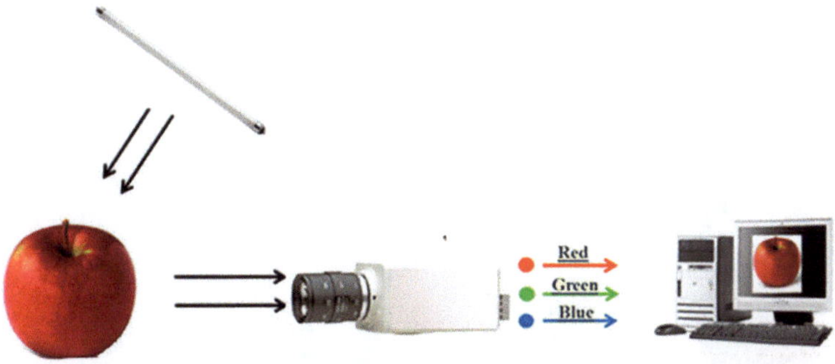

Fig. 4.2 A basic computer vision system schema

light (light source), the interaction of light with materials and the physiology of human vision.

It is known that a light source comprises any process able to emit, re-emit, or conduct energy in sufficient quantities to produce light, in other words, produces electromagnetic radiation whose wavelengths are inside the visible region. There are basically two types of light sources: natural light, such as the sun, and artificial light, which can be based on different types of technologies, such as incandescent lamps, fluorescent lamps, LEDs, among others. The distribution of the electromagnetic radiation emitted by these lamps depends on the emission material and the conditions of its use.

The electromagnetic radiation originating from the light sources interacts with various types of materials. Each material has its own characteristics due to their molecular formation, thus presenting different properties when it interacts with light. The result is the material color, which depends on its pigment constituents.

Therefore, a beam of radiation that is emitted can undergo reflection, transmission, or absorption when it focuses on a particular material. It is noteworthy that every material absorbs radiation in some part of the electromagnetic spectrum and the quantity of energy absorbed relies on the selectivity of the absorbing material (pigment) and the wavelength of the radiation.

Thus, one can consider two different types of color: the color originating from a light source and the color pigment originating from a particular material, making the task of specifying color impossible without including the knowledge of the radiation source, since it provides the light that is reflected by the material that, in turn, causes the color perception in the human eye, as shown by Fig. 4.3.

The electromagnetic radiation emitted from light sources or from objects due to reflection or transmission is perceived by the human vision system with the photosensitive cells present in the retina, as seen above in Fig. 4.3. The physiology of the human eye states that the retina is composed of two types of receptor cells that, due to their shapes, are called rods and cones. The cones are found primarily in the center of the retina and are responsible for vision with higher brightness, known as photopic vision. They are also responsible for color vision, being

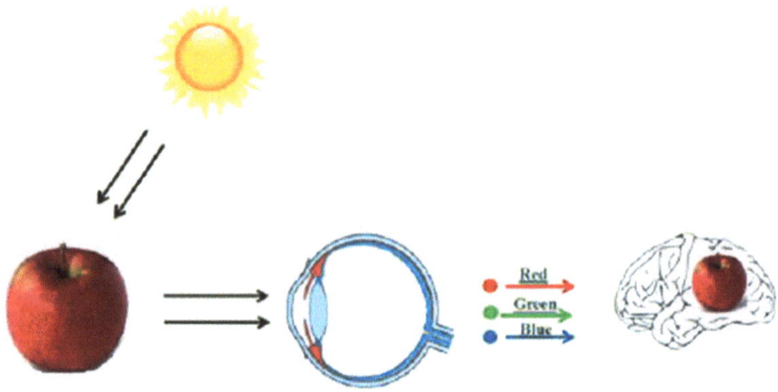

Fig. 4.3 A basic human vision system schema

sensitive to long, medium, and short wavelengths (red, green, and blue), commonly known by their initials in English R, G, and B. The rods are located at the outer edges of the retina and are responsible for vision in low light, being sensitive only to shades of gray.

It can be concluded that the phenomenon of color depends on the nature of light (light source), the interaction of light with materials, and the physiology of human vision. In summary, the phenomenon of color perception is based on our body's response to stimuli received from an electromagnetic radiation originated from a light source, reflected or not for a given material and focusing on our eyes.

To standardize the identification of colors it is necessary to define some characteristics of the light sources in order to standardize operating conditions. One is the color temperature (CT), which expresses the appearance of the color of the light emitted by the light source. Its unit of measurement is the Kelvin (K). The higher the CT, the brighter the hue of a light's color. The correlated color temperature (CCT) is defined as the temperature of the Planck radiator whose perceived radiation more closely resembles that of a given stimulus with the same brightness under specific observation conditions [4].

4.4 Color Analysis Using Image

The perception of an object (or of the acquired image of an object) is a function of the spectral property of this object, the lighting, and its acquisition (human or not). It is known that the phenomenon of color depends on the nature of light (light source), the interaction of light with materials, and the physiology of human vision (or image acquisition device). The phenomenon of color perception is based on the body's response to stimuli received from an electromagnetic radiation originated from a light source, reflected or not on a given material, focusing in the viewer's eyes. Likewise, the acquired image of an object is a function of the spectral

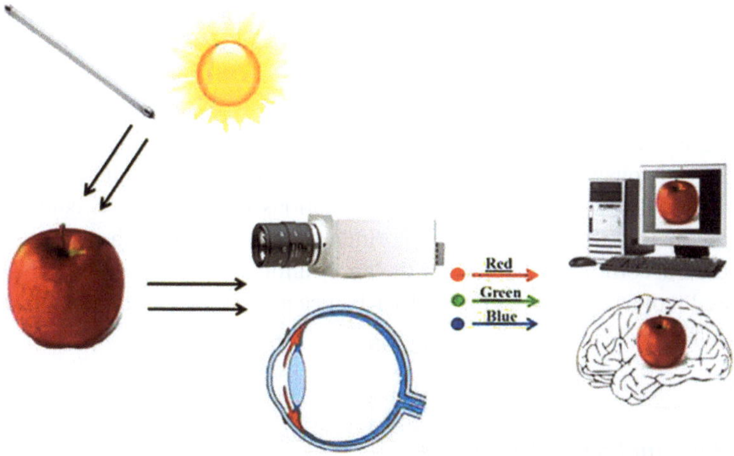

Fig. 4.4 Computer vision system × human vision system

property of the object, lighting, and camera in response to the stimulus received, as compared by Fig. 4.4.

Just like the human eye, a digital camera has sensors that convert radiation (stimulus) into electrical charges, which are then digitalized. The image sensors used in most digital cameras are charge coupled device (CCD) or complementary metal oxide semiconductor (CMOS). Each sensor is composed of two filters, one black and white with sensitivity to different intensities of radiation and another sensitive to red, green, and blue, called the Bayer filter, deployed in order to make the acquisition of radiation by the camera as similar as possible to the human eye.

Comparing the spectral distributions of filters used in some models of cameras, one can observe a difference between models and manufacturers. Comparing the spectral distribution of the cameras with the distribution of color equalization functions of *Commission Internationale de l'Eclairage*—the International Commission on Illumination (CIE) there is a big difference between them, technically demonstrating the need for calibration and characterization of these when used in quantitative automatic inspection systems.

The process of image digitalization is generated from the mapping of an analog image, in which the latter is divided into several points, forming a two-dimensional array $f(x, y)$. Each point of this matrix is called the pixel of the image, and, due to the color filter used, these two elements are connected with the spatial resolution of the image's color, which is typically represented by color systems to enable their representation. Similarly to the human eye, the color system that is most similar to the image acquisition system is the RGB due to the formation of the three additive colors red, green, and blue.

Various features can be extracted from the RGB values obtained during the processing step, including some appearance characteristics such as the color. Other color systems are also used in the analysis of color of an image from the conversion of RGB values in other systems, such as HSL, HSI, HSB, and HSV that are widely employed since they consider concepts of hue, saturation, and brightness.

For calibration of digital cameras it is necessary to convert the different color systems used into a color system that is known to be subsequently compared with standards previously calibrated in order to give the system the necessary traceability. The color system used in such calibrations is the CIEXYZ color system, developed by the CIE to be used in colorimetry and, from this, other color systems can be used according to the need of the application area. In the industrial area the most used color system is the CIELAB because it is the closest to human perception.

4.4.1 Parameters Influence

Because an object's color is one of the main indicators of perception of the products quality, the factors that influence color analysis should be considered in the system of visual assessment.

The image acquisition step is a physical process that depends on a large number of parameters for obtaining a quality final image. Some parameters influence the image quality to be obtained, such as the device used to acquire the image, the lighting system used, and the background color. There are many types of devices used in image acquisition, such as digital cameras, scanners, video cameras, and thermal. The most commonly used devices are digital cameras, which cover about 58 % of the total of devices [1]. Another important parameter in the image acquisition device is a detection system, which can be CCD or CMOS, since each has differences in the speed of image acquisition and sensitivity to light, both of which have advantages and disadvantages that must be studied case by case.

The type of lighting is also of influence in the resulting picture. It is important to note that some features of the light source used, such as CCT and color rendering index (CRI), are decisive in the final perception of the image and should be set according to the research objective [5]. There are many light sources available, which have quite different properties between each other. Thus, to standardize the identification of colors it is necessary to define some characteristics for the light sources, such as the CRI and CCT [4]. With the knowledge of the Spectral Power Distribution (SPD) it is possible to calculate the CCT that expresses the color appearance of the light emitted by the source. The CCT increases from red to blue; the higher its value is, the brighter will be the color hue of the light, and the lower, the more yellowish, thus influencing in the perception of the color of the material. Lamps with CCT below 3,100 K are considered "hot", and those with values above 4,100 K are considered "cold" while those in the range of 3,100–4,100 K, are considered neutral color or moderate (Table 4.1).

The CRI is also an important parameter, quantifying the fidelity with which the colors are reproduced when struck by a particular source [6]. The CRI is calculated

Table 4.1 Examples of SPD of some light sources [9, 10]

Source	Spectrum	CCT (K)	CRI
Fluorescent	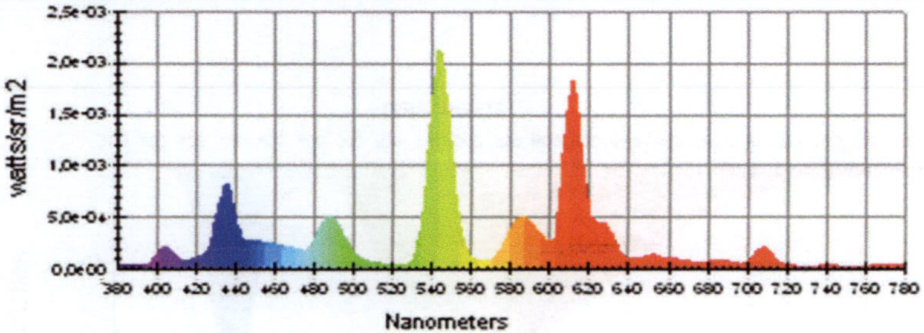	3,834 ($U_{CCT} = 47$ K)	60.2 ($U_{CRI} = 2.7$)
High pressure sodium vapor	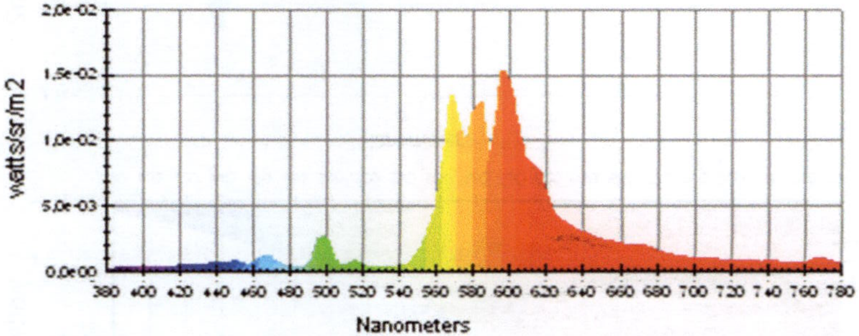	2,059 ($U_{CCT} = 13$ K)	22.4 ($U_{CRI} = 1.9$)

(continued)

Table 4.1 (continued)

Source	Spectrum	CCT (K)	CRI
Incandescent	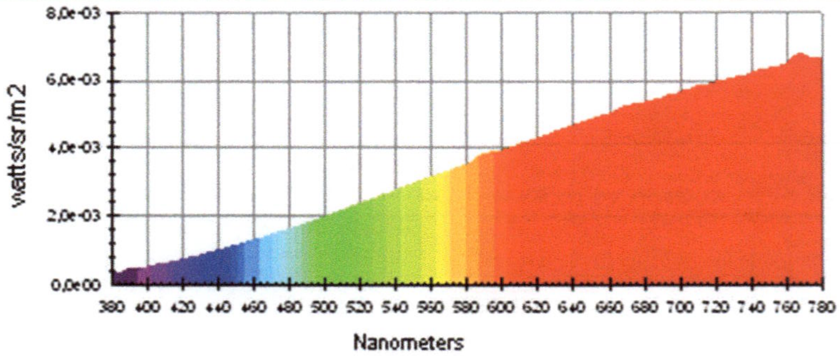	2,856 (U_{CCT} = 21 K)	99.8 (U_{CRI} = 2.3)
LED	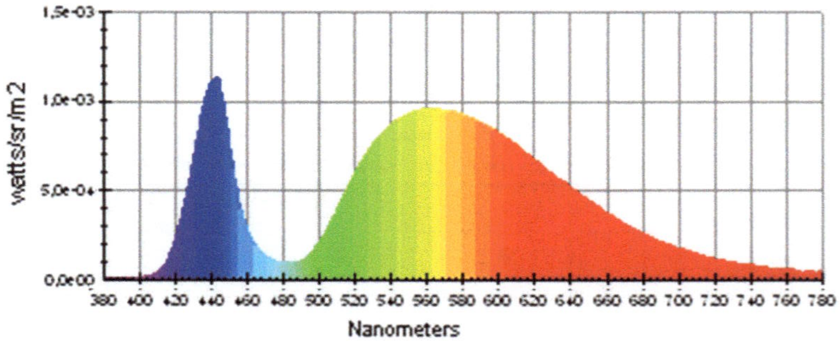	4,284 (U_{CCT} = 69 K)	67.2 (U_{CRI} = 1.8)

from the SPD of the source, and is an average calculated from the difference in color of a prespecified set of samples under both a reference illuminant and a test illuminant [7]. The best performance of a light source, when compared to a given reference, generates a CRI of 100. A CRI greater than 80 is generally acceptable depending on the requirement of the application that a lamp must meet [8] (Table 4.1).

The types of illumination most commonly used are fluorescent lamps (65 %), incandescent bulbs (10 %), natural lighting (19 %), and LED (2 %) [1]. Undoubtedly natural lighting would be the best option; however, its characteristics widely changes depending on the sun angle, time of day, weather conditions, etc. Incandescent lamps have a high CRI value, but have high energy consumption and its usage has been avoided. The technology of fluorescent lamps has replaced incandescent bulbs, but their characteristics are varied depending on the composition (amount or type of phosphorus) and it also contains mercury. Even so, its consumption has greatly increased, especially with the use of compact fluorescent lamps, due to their low power consumption and reasonable lifetime.

So, the lighting used to characterize a material must be selected regarding the properties of the source (lamp) and its application in particular cases requires experimental testing [9].

Mohan et al. [11] determined in a comparative study that the reflectance of grains in the near infrared region allowed for a better classification accuracy than the reflectance in the visible region, thus enabling the choice of the best source to be used in the classification of grains. Manickavasagan et al. [12] studied the influence of three types of lighting in the classification of grains: incandescent with 2,870 K, fluorescent ring with 3,000 K and fluorescent tubular with 4,100 K. They both concluded that fluorescent lamps of the T8-type, normally used in classrooms, have better accuracy for identifying defects in grains compared to halogen bulbs, because these have a greater emission of radiation in the infrared region, thus confirming the study by Mohan et al. [11].

The background color has a direct influence on the analysis of color, being another important parameter for image acquisition. Research demonstrates that the background where the material is analyzed can influence the perception of the color of the material [13–15]. According to Brown and MacLeod [13], the perceived color of a scene relies on the ratio between the light signals from that point and light signals from surrounding areas of this scene.

Because the appearance of objects influences directly the consumer's decision about the quality of the product, Dobrzański Rybczynski [16] studied the influence of the packaging of fruits and vegetables in color perception. In this study, oranges, carrots, beetroots, and parsleys were packed with nets of different colors in order to analyze their influence on color and consumer preference. It was concluded that the red colored net influences the color of the orange, appearing to be a more mature fruit, influencing the final consumer.

Black or white backgrounds are the most used [17] because, for the purpose of the analysis, they facilitate the segmentation stage. Blasco et al. [18] presented a study varying the background color in the selection of pomegranate seeds, concluding that the blue background contributed to a better segmentation.

The diversity of industrial products (which have a variety of sizes, shapes, textures, and color) emphasizes the importance of defining standardized parameters during the classification and packing [1], since inadequate illumination may hinder the identification of defects and the color of the products analyzed [9] and the lack of standardization of the background color used for the classification of products (usually conveyor belts or benches) can impair selection process.

4.5 Case Study

Gomes et al. [19] developed a research in order to define a new method for nondestructive testing using colorimetric techniques and computer vision for characterizing color using digital images applied to integrated fruit production, focusing on standardization of measurements, considering the factors of influence. In the initial study, the banana was chosen as a case study in the development of this research, and as a result, a new methodology was developed to characterize the stages of maturation of banana using colorimetric analysis, proposing a standard for the industry [10].

Figure 4.5 shows an example of visual assessment in monitoring of the banana ripening for selection and fruit trade [10] using a halogen lamp as source (CCT = 2,856 K, CRI = 99.8) and black background.

Figure 4.6 shows the difference of perception in an image when using different types of illumination. Figure 4.7 shows the difference in perception of the image when different background colors are used [10].

Large color differences were found when comparing the same ripening subclass using different sources. Such differences justify the need for greater concern about the lighting system employed in the classification area. When comparing different background colors, red and blue backgrounds offered most influence on the evaluation of ripening subclasses.

Therefore, one should evaluate the characteristics of interest to define the parameters to be used in image acquisition in order to get a better image and better accuracy in measuring color from the image. For improving the results it is suggested the following steps for defining an image acquisition system for color analysis: a study of the sample; a study of the best background color; a study of the best light source; and finally the calibration of the measurement system under these conditions (Fig. 4.8).

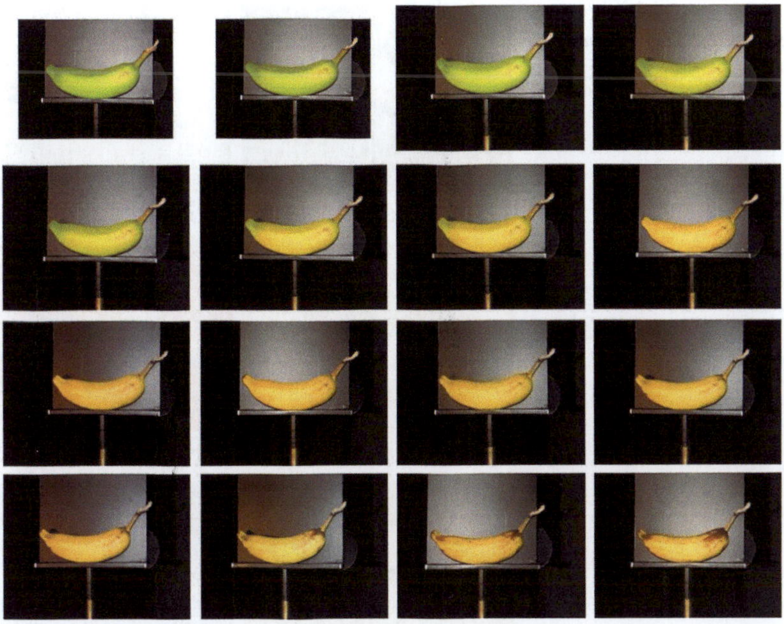

Fig. 4.5 Daily evolution of the ripening of banana 'Prata'

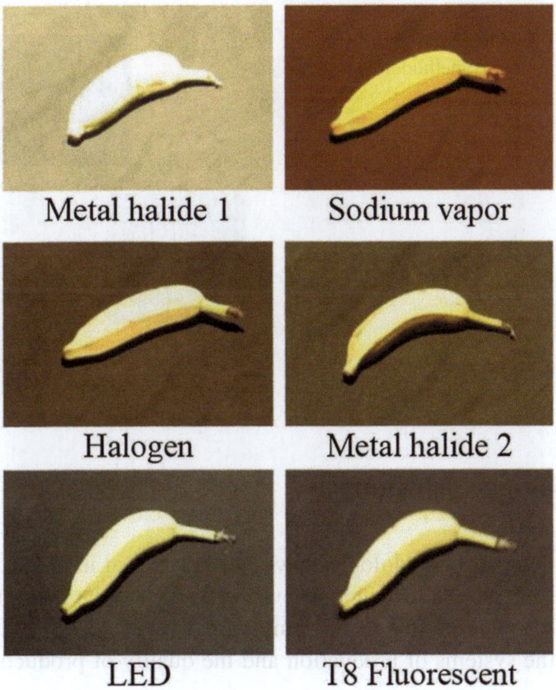

Fig. 4.6 Fruit analyzed under different *lighting* [10]

Fig. 4.7 Fruit analyzed under different *background colors*

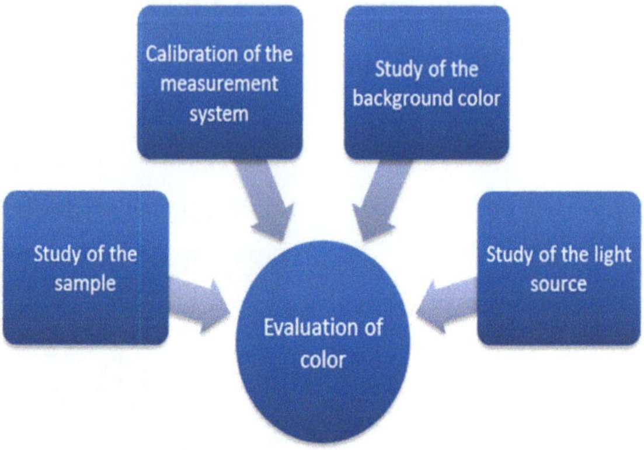

Fig. 4.8 Summary of the proposed methodology

4.6 Conclusions

Several methodologies have been proposed in order to characterize different types of materials for computer vision in so that the inspection process is optimized and losses are minimized. Many works have been developed to increasingly improve the systems of production and the quality of products. In general, most CVS have well-defined five steps. But there is no agreement in what are the right tools and strategies for image acquisition and processing. Each of the technological options presented have their particular response. However, there is an emergent need to develop such methodologies in order to meet a great ongoing demand and cover all the requirements and restrictions of existing procedures due to the specificity of each product, also aiming to standardize these new technologies.

Acknowledgements The authors would like to acknowledge FAPERJ (under grants E-26/103.591/2012, E-26/103.618/2012 and E-26/171.362/2001) for its financial support. The authors would also like to acknowledge their colleagues from UFF and Inmetro for the support while conducting the experiments. They also acknowledge Dr. Ana Paula Dornelles Alvarenga and Marcelo Bezerra Guedes for the technical discussions.

References

1. Gomes, J.F.S., Leta, F.R.: Applications of computer vision techniques in the agriculture and food industry: a review. Eur. Food Res. Technol. **235**, 989–1000 (2012)
2. Zhendong, L., Haiye, Y., Hongnan, L., Hongxia, Z.: The application study on building materials with computer color quantification system. In: Proceeding of SPIE, vol. 6033, 603307-1 (2005)
3. Conci, A., Azevedo, E., Leta, F.R.: Computação Gráfica—Teoria e Prática, v 2. Campus Elsevier Ed., Rio de Janeiro (2008)
4. CIE 15.3: Colorimetry. CIE Publication, Vienna (2004)
5. Ling, Y., Vurro, M., Hurlbert, A.: Surface chromaticity distributions of natural objects under changing illumination. In: Proceeding of the 4th European Conference on Colour in Graphics, Imaging and Vision (CGIV), pp. 263–267 (2008)
6. CIE 13.3: Method of Measuring and Specifying Colour Rendering Properties of Light Sources. CIE Publication, Vienna (1995)
7. Luo, M.R.: The quality of light sources. Color. Technol. **127**, 75–87 (2011)
8. LRC—Lighting Research Center: Recommendations for Specifying Color Properties of Light Sources for Retail Merchandising. Alliance for Solid-State Illumination Systems and Technologies, vol. 8, issue 2 (2010)
9. Gomes, J.F.S., Vieira, R.R., Oliveira, I.A.A., Leta, F.R.: Influence of illumination on the characterization of banana Ripening. J. Food Eng. **120**, 215–222 (2014)
10. Gomes, J.F.S.: Padronização de metodologia para caracterização de cor por imagem aplicada à seleção de frutas. Doutoral Tese from Universidade Federal Fluminense (2013)
11. Mohan, L.A., Karunakaran, C., Jayas, D.S., White, N.D.G.: Classification of bulk cereals using visible and NIR reflectance characteristics. Can. Biosyst. Eng. **47**, 7.7–7.14 (2005)
12. Manickavasagan, A., Sathya, G., Jayas, D.S.: Comparison of illuminations to identify wheat classes using monochrome images. Comput. Electron. Agric. **63**, 237–244 (2008)
13. Brown, R.O., MacLeod, D.I.A.: Color appearance depends on the variance of surround colors. Curr. Biol. **7**, 844–849 (1997)
14. Shevell, S.K., Wei, J.: Chromatic induction: border contrast or adaptation to surrounding light? Vision. Res. **38**, 1561–1566 (1998)
15. Sánchez-Zapata, E., Fuentes-Zaragoza, E., Vera, C.N.R., Sayas, E., Sendra, E., Fernández-López, J., Pérez-Alvarez, J.A.: Effects of tuna pâté thickness and background on CIEL*a*b* color parameters and reflectance spectra. Food Control **22**, 1226–1232 (2011)
16. Dobrzański Jr, B., Rybczyński, R.: Influence of packing method on colour perception improving the appearance of fruits and vegetables. Res. Agric. Eng. **54**(2), 97–103 (2008)

17. Meléndez-Martínez, A.J., Vicario, I.M., Heredia, F.J.: Correlation between visual and instrumental colour measurements of orange juice dilutions: effect of the background. Food Qual. Prefer. **16**, 471–478 (2005)
18. Blasco, J., Cubero-García, S., Alegre-Sosa, S., Gómez-Sanchís, J., López-Rubira, V., Moltó, E.: Automatic inspection of the pomegranate (Punica granatum L.) arils quality by means of computer vision. Span. J. Agric. Eng. **6**(1), 12–16 (2008)
19. Gomes, J.F.S., Vieira, R.R., Leta, F.R.: Colorimetric indicator for classification of bananas during ripening. Sci. Hortic. (2013). doi:10.1016/j.scienta.2012.11.014

Part II
Medicine Applications

Chapter 5
An Automated System for 3D Segmentation of CT Angiograms

Y. Wang and P. Liatsis

Abstract This chapter presents a novel automated two-step algorithm for segmentation of the entire arterial tree in 3D contrast-enhanced Computed Tomography Angiography (CTA) datasets. In the first stage of the proposed algorithm, the main branches of the coronary arteries are extracted from the volume datasets based on a generalised active contour model by utilising both local and global intensity features. The use of local regional information allows for accommodating uneven brightness distribution across the image. The global energy term, derived from the histogram distribution of the input images, is used to deform the contour towards to desired boundaries without being trapped in local stationary points. Possible outliers, such as kissing vessel artefacts, are removed in the following stage by the proposed slice-by-slice correction algorithm. Experimental results on real clinical datasets have shown that our method is able to extract the major branches of the coronaries with an average distance of 0.7 voxels to the manually defined reference data. Furthermore, in the presence of kissing vessel artefacts, the outer surface of the coronary tree extracted by the proposed system is smooth and contains less erroneous segmentation as compared to the initial segmentation.

Keywords Computed tomography angiography · 3D segmentation · Coronary arteries · Active contour models

---

Y. Wang
College of Astronautics, Nanjing University of Aeronautics and Astronautics, Nanjing, China
e-mail: yinwangee@nuaa.edu.cn

P. Liatsis (✉)
Department of Electrical and Electronic Engineering, City University London, London, United Kingdom
e-mail: P.Liatsis@city.ac.uk

F. R. Leta (ed.), *Visual Computing*, Augmented Vision and Reality 4,
DOI: 10.1007/978-3-642-55131-4_5, © Springer-Verlag Berlin Heidelberg 2014

List of Abbreviations

CTA	Computed tomography angiography
CAD	Coronary artery disease
CT	Computed tomography
EM	Expectation maximisation
GMM	Gaussian mixture model
LAD	Left anterior descending
LCA	Left coronary artery
LCX	Left circumflex
LM	Left main coronary
RCA	Right coronary artery
WHO	World Health Organization

5.1 Introduction

According to the most recent statistics by the World Health Organization (WHO), coronary artery disease (CAD) is one of the most prevalent causes of death in the world. Approximately 17 millions people are killed due to one or more types of CAD every year worldwide [1]. Atherosclerosis is a common cause of CAD, where plaque builds up on the insides of arteries. Plaques are typically made up of fat, cholesterol, calcium and other substances found in the blood. If left untreated, it will harden and narrow the arteries over a period of years. As a consequence, the flow of oxygen-rich blood to organs and other parts of body will be reduced thus leading to serious problems, including myocardial/cerebral infarction, or even death. Figure 5.1 shows a comparison before and after plaque builds up within a blood vessel.

Coronary arteries are the network of blood vessels which supply blood to the myocardium. CAD occurs when the coronaries are occluded and become constricted, which makes the heart become starved for oxygen and other nutrients and eventually stop beating. As illustrated in Fig. 5.2, the arterial tree basically consists of two main branches, namely the Left Coronary Artery (LCA) and Right Coronary Artery (RCA), which arise from the root of the aorta, near the top of the heart. In LCA, the initial segment between the aorta and the first bifurcation is called the left main (LM) coronary. The LM typically branches into the left anterior descending (LAD) and the left circumflex (LCX) arteries. On the other hand, the RCA normally originates from the right coronary cusp and travels to the posterior interventricular branch. In 85 % of cases, the RCA is the dominant vessel and supplies the posterior descending branch which travels in the PIV groove. However, the exact anatomy of the coronary arteries could exhibit large differences from individual to individual [2].

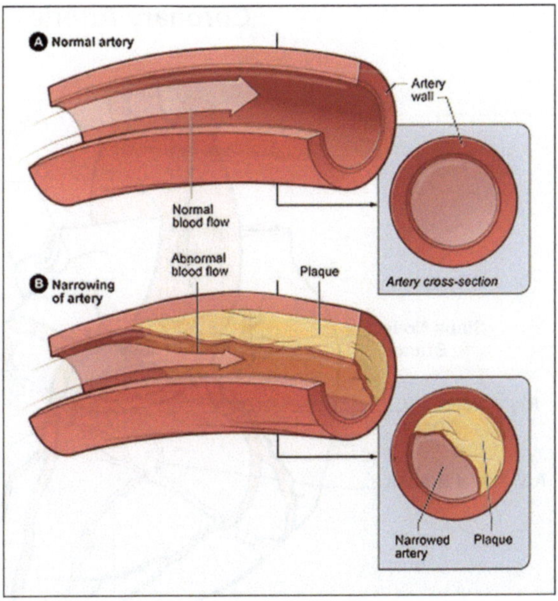

Fig. 5.1 Effect of plaque builds up. **a** Shows a normal artery with normal blood flow. **b** Shows an artery with plaque buildup [2]

In order to evaluate the development of coronary artery disease, efficient screening procedures are urgently needed for early diagnosis and ultimately timely treatment. From the clinician's point of view, it is essential to detect and visualise the branches of the arteries, which may be associated to stenosis (narrowing of the arteries). Then, potentially carrying out haemodynamic studies to establish the risk of a patient developing a heart attack, followed by the application of appropriate treatment based on medication or surgical procedures, such as diagnostic cardiac catheterisation and stent implantation [3, 4].

Recent advances in vascular imaging technology using multiple-slice Computed Tomography (CT) provide volumetric datasets with high spatial and temporal resolution. This has given rise to CT becoming a frequently used and reliable imaging modality for the early diagnosis and quantification of atherosclerosis in clinical practice. However, clinical diagnosis of CAD by means of CT imaging is difficult because of the huge amount of data produced in the cardiac CT scanning process (typically, more than 300 slices of axial CT images are produced for a patient). Interpretation of a CT study is a labour-intensive and time-consuming task even with the assistance of semi-automatic software packages [5], as the radiologist has to track and carefully examine each branch of the arteries, segment by segment. Therefore, automated algorithms for segmentation of the coronaries in CT angiograms will facilitate and enhance the accuracy and reliability of the clinical evaluation.

In the following section, we provide an overview of recent developments in the field of the research. Next, we present the proposed framework in Sect. 5.3. This is followed by the presentation of results demonstrating the benefits of the proposed approach. Finally, the conclusions and future developments of this research are presented.

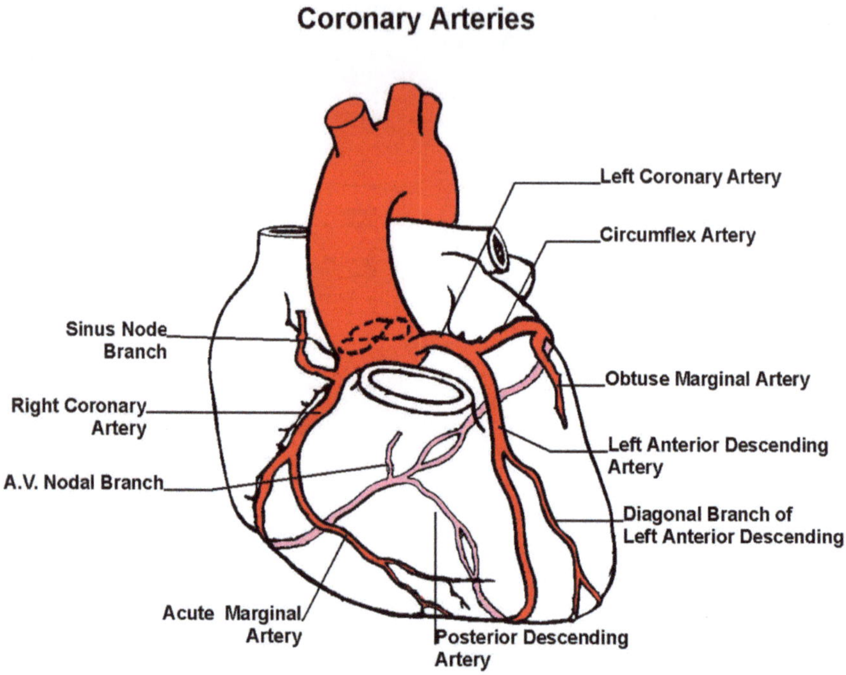

Fig. 5.2 Illustration of the coronary arteries of human, including: the right coronary artery (RCA), left anterior descending (LAD) and left circumflex (LCX)

5.2 State of the Arts in Vessel Segmentation

Despite numerous past and on-going research efforts in the past decade, segmentation of vascular structures from CT images remains a challenging topic due to the small size of vessels in medical images and the complex pathology. Since their introduction as a means of front propagation-based segmentation method, active contour models (also known as snakes) received a great amount of attention by the medical image processing community [6–12]. Active contour models for image segmentation iteratively deform a contour in order to minimise a user-defined energy functional, which often depends on the shape of the contour and its position in the image. Such methods are usually implemented using level sets [13], where the boundaries of the object to be segmented are embedded as the zero level of a higher dimensional level set function. Due to their ability to deal with topological changes, such as region merging and splitting, level sets-based active contour models are usually employed in segmentation of the vascular structures in medical images. Active contour-based methods can be categorised in two folders, when considering image-driven energy: edge-based and region-based models.

In the early edge-based models [11, 14, 15], the active contour deforms with a speed $\boldsymbol{F}$ based on the derivatives of the image, which approaches zero at high image gradients. These methods make use of the local edge information to stop curve deformation, which makes them robust to region inhomogeneities. On the other hand, they are sensitive to image noise and contour initialization due to the use of local gradients. Yang and her colleagues proposed an improved algorithm to segment coronary arteries based on a Bayesian probabilistic framework [16]. In their work, the image force is redefined using posterior probabilities, calculated from the global histogram distributions, to more accurately terminate the surfaces at desired boundaries. In their method, the posterior probabilities are obtained from global statistics, which cannot handle the varying brightness and contrast changes over the image. Thus, their method is not capable of segmenting small and distal segments of the coronary arteries due to their relatively low intensity contrast. In medical images, the borders between different objects are not always clearly defined by the gradients, and thus, the contour may leak into adjacent regions, when using the edge function as the stopping criterion. Nain et al. [17] incorporate a soft shape prior into the conventional active contour model. They propose the application of a shape filter to locally describe the shape of the segmented region. As illustrated in Fig. 5.3, the shape filter is defined as a ball structure centred on each point along the contour with radius r. This measures the percentage of voxels belonging to both the ball and the object (i.e. the regions inside the contour). The output of the shape filter is high when the current point belongs to a region corresponding to leakage. Conversely, lower values of the filter's output indicate that the current point is within the vessel. The filter response then serves as the external energy of the active contour, penalising leakages during the curve evolution. However, the shape filter cannot discriminate vessel bifurcations from leakage areas, and may result in undesired gaps in the vicinity of vessel bifurcations.

Region-based image segmentation methods, which utilise intensity information obtained from image regions, are more robust to image noise. In these methods, region statistics along the contour are calculated to drive the segmentation process. Under the assumption that the object and the background are approximately uniformly distributed, Chan and Vese [6] proposed an active contour model using regional statistics to segment the object of interest in two-phase images. Their work was later extended to multiple-phase images [18], where the N regions (phases) are represented by $\log_2 N$ level set functions. However, empty regions will be produced when less than N regions are present in the image. To handle more complex intensity distributions, non-parametric method is applied to estimate regional statistics [19]. The aforementioned methods, however, solely based on global intensity statistics, are inefficient in cases where regional statistics are spatially varying across the image. Localised approaches [20–22], where regional statistics are calculated in a neighbourhood of the active contour, have recently emerged to overcome this problem. Such models are more robust to local variations of the region of interest and therefore improve the overall segmentation results. However, segmentation based on local decisions alone may not be sufficient to drive the contour to stop at the desired boundaries, since the contour may be trapped in undesired local stationary points. Moreover, the selection of appropriate scales also poses additional difficulties.

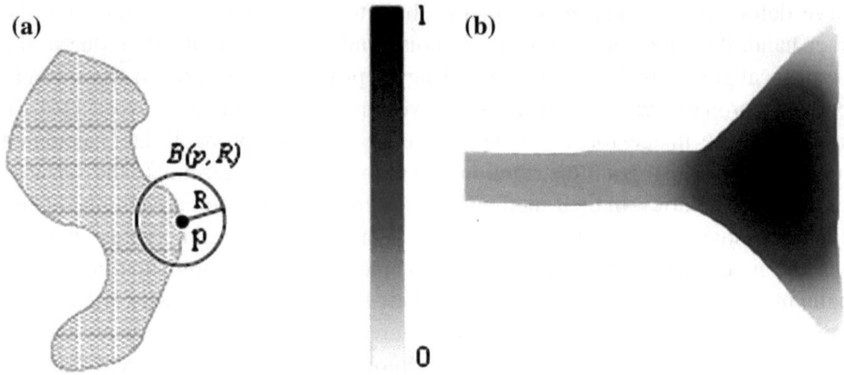

Fig. 5.3 Graphic demonstrating the effect for the shape filter. **a** Shows of the shape filter centred at x and with the radius R. **b** The output of the shape filter [17]

Another issue in the segmentation of vascular structures in coronary CT images is the so called 'kissing vessel artefacts', i.e. the non-coronary vessels in close proximity to the arteries. This is a partial volume problem which is often encountered in CT angiograms [23]. It may result to artificial vessel junctions and thus distort the geometry of the vessels. To address this problem, Wong and Chung [24] proposed the tracing of vessel centrelines and segmentation of their cross-sections based on a probabilistic vessel-axis tracing framework. Their algorithm allows user's interaction to produce the desired traces through the abnormal regions, which contain the kissing vessels, lesion vessels (e.g. the stenosis) and vessel junctions. The final segmentation in these regions is then derived from the axis of the cross-sections.

5.3 Proposed Framework

This section presents the proposed two-step system for segmentation of coronary arteries in CTA images by using both global and local intensity statistics. We deal with the varying image brightness characteristics by computing regional statistics locally, in the neighbourhood of the active contour. The global intensity constraint, on the other hand, is utilised to evolve the contour to the desired boundaries without being trapped in local minima. The proposed approach is implemented in a Bayesian probabilistic framework to incorporate these two homogeneity constraints. Possible outliers, such as kissing vessel artefacts, are removed in the subsequent stage by a novel slice-by-slice correction scheme, which uses multiple regions competition.

5.3.1 Coronary Arteries Segmentation

We commence our analysis by assuming that voxels in contrast-enhanced CTA images fall into three categories, i.e. the air in the lungs, soft tissues and blood-filled regions. Then, we use a Gaussian Mixture Model (GMM) to fit the histogram of the input CTA volume to estimate the probability density function for each class, as shown in Fig. 5.4a. The mean and variance for each class are estimated using the Expectation–Maximization (EM) method. We use prior anatomical knowledge that coronaries are located on the outer surface of the heart, and thus, we neglect the class corresponds to the air to obtain a bi-modal histogram (see Fig. 5.4b). The first peak (T_1) in the fitted histogram corresponds to soft tissues in the heart, which reflect the intensity distribution of the background pixels. According to the assumption that voxels with intensity values less than T_1 as belonging to the background, while voxels with intensity values greater than this threshold are treated as potential objects of interest (i.e. blood-filled regions), we assign each voxel in the volumetric data with a fuzzy label, which indicates the probability of the voxel belonging to the object.

In this research, we formulate the labelling function as a normalised cumulative density function of the histogram. We normalise the labelling function between -1 and 0 for voxels with intensity values between 0 and T_1, and the output of the labelling function bounded between 0 and 1 for the input voxels with intensity values greater than T_1. Thus, the function is defined as follows:

$$L(x) = \begin{cases} [-1, 0), & \text{if } x \text{ belongs to the background} \\ [0, 1], & \text{if } x \text{ is a potential 'object'} \end{cases} \tag{5.1}$$

Let $\Omega_{\boldsymbol{x}}$ denote a neighbourhood with a radius r centred at $\boldsymbol{x}$ on the active contour $C(\boldsymbol{x})$.The localised image, $\Omega_{\boldsymbol{x}}$, can be partitioned into two sub-regions by the active contour, i.e. the regions inside and outside the active contour, respectively. Hence, we define the probability of a voxel being classified as belonging to the region Ω_i as follows:

$$P_i = P(I(\boldsymbol{y})|\boldsymbol{y} \in \Omega_i \cap \Omega_{\boldsymbol{x}}) = \frac{1}{\sqrt{2\pi}\sigma_i} \exp\left(-\frac{(\mu_i - I(\boldsymbol{y}))^2}{2\sigma_i^2}\right) \tag{5.2}$$

where $\{\Omega_i | i = 1, 2\}$ denote the regions inside and outside the contour. $I(\boldsymbol{y})$ is the image intensity at $\boldsymbol{y}$, μ_i and σ_i represent the mean and the variance derived from region Ω_i, respectively. Note that, we use $\boldsymbol{x}$ and $\boldsymbol{y}$ as two independent spatial variables to represent a single point in the image domain. Let $C(\boldsymbol{x})$ denotes a contour, representing the boundary of the object to be segmented. For each point along the contour, given its local image $\Omega_{\boldsymbol{x}}$ and the labelling function $L(\boldsymbol{y})$, the posterior probability of a voxel $\boldsymbol{y}$ being classified as belonging to the sub-region $\Omega_i \cap \Omega_{\boldsymbol{x}}$ can be defined as:

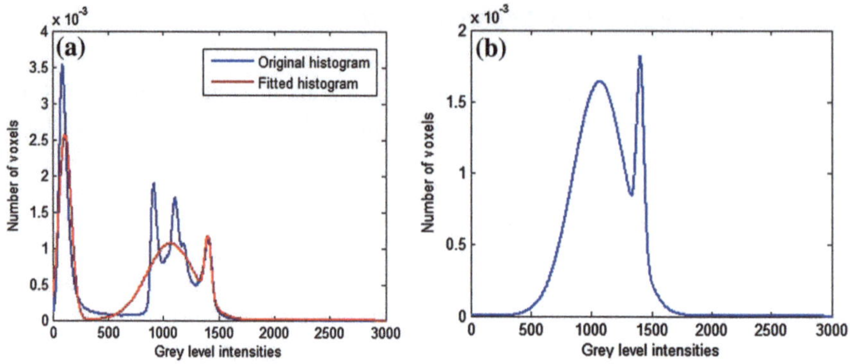

Fig. 5.4 The histogram of the CTA image. **a** The histogram and the fitted mixture model of the CTA image. **b** The bi-modal histogram obtained within the heart region

$$P(y \in \Omega_i \cap \Omega_x | I(y), L(y)) = \frac{P(I(y), L(y) | y \in \Omega_i \cap \Omega_x) P(y \in \Omega_i \cap \Omega_x)}{P(I(y), L(y))} \quad (5.3)$$

where $P(y \in \Omega_i \cap \Omega_x\ i = 1{,}2)$ is the prior probability of the current voxel being assigned to region Ω_i among all the possible partitions within the local image Ω_x. This term can be ignored, if equal probabilities are assumed for all partitions of the image. $P(I(y), L(y))$ denotes the joint probability density distribution of the grey level value $I(y)$ and the labelling function $L(y)$, which is independent of the segmentation of the image and can therefore be neglected. We assume that the voxel labels and the grey level intensity distribution are independent. The posterior probability for each voxel can thus be computed as:

$$P(I(y), L(y) | y \in \Omega_i \cap \Omega_x) = P(I(y) | y \in \Omega_i \cap \Omega_x) P(L(y) | y \in \Omega_i \cap \Omega_x) \quad (5.4)$$

The prior probability of $P(I(y)|y \in \Omega_i \cap \Omega_x)$ has been already defined in Eq. 5.2. In order to compute the posterior probabilities in Eq. 5.4, the prior probability of the labelling function should be known. In this research, we model the prior probability distribution of the labels as:

$$P(L(y) | y \in \Omega_i \cap \Omega_x) \propto \exp\left(\frac{v}{2} L(x) R(x) k_\sigma(x, y)\right) \quad (5.5)$$

where:

$$k_\sigma(x, y) = \frac{1}{\sqrt{2\pi}\sigma} \exp\left(-\frac{(x - y)^2}{2\sigma^2}\right) \quad (5.6)$$

represents the weighting kernel, which is a decaying function of the distance between x and y. v_2 is the overall weight that determines the influence of the labels on the segmentation. $R(x)$ is a normalised Boolean function indicating whether the current voxel $\boldsymbol{x}$ is located inside the contour $C(\boldsymbol{x})$ within the local image $\Omega_{\boldsymbol{x}}$. According to Eq. 5.5, when a voxel, located at position $\boldsymbol{y}$, is classified as belonging to the object, i.e. $R(\boldsymbol{x}) = 1$, then a point $\boldsymbol{x}$ in its vicinity has a high probability of being marked as the object (i.e. $L(\boldsymbol{y}) \to 1$) and is less likely to be considered as the background ($L(\boldsymbol{y}) \to -1$).

Maximising the posterior distribution in Eq. 5.4 is equivalent to minimising its negative logarithm, and thus the energy function can be presented as:

$$\begin{aligned} E = &- \int H'(\varphi(\mathbf{x})) \sum_{i=1}^{2} \{ \int_{\Omega_i \cap \Omega_x} [\log P(I(\mathbf{y})|\mathbf{y} \in \Omega_i \cap \Omega_{\mathbf{x}}) \\ &+ \log P(L(\mathbf{y})|\mathbf{y} \in \Omega_i \cap \Omega_{\mathbf{x}}) dy] M_i(\varphi(\mathbf{y}))\} dx \\ &+ \mu \int |\nabla H(\varphi(\mathbf{x}))| dx \end{aligned} \tag{5.7}$$

where $M_1(\phi(\boldsymbol{y})) = H(\phi)$ and $M_2(\phi(\boldsymbol{y})) = 1-H(\phi)$, and $H'(\cdot)$ denotes the derivative of Heaviside function:

$$H(\phi) = \begin{cases} 0, & \text{if } \phi \leq 0 \\ 1, & \text{if } \phi > 0 \end{cases}$$

and the associated Euler–Lagrange equation can be then obtained as:

$$\frac{\partial \phi}{\partial t} = \delta(\phi) \left(\mu \operatorname{div}\left(\frac{\nabla \phi}{|\nabla \phi|} \right) + \log \frac{p_1}{p_2} + v \int L(\boldsymbol{x}) k_\sigma(\boldsymbol{x}, \boldsymbol{y}) \mathrm{d}\boldsymbol{y} \right)$$

$$\begin{aligned} p_1 &= \int\limits_{\Omega_1 \cap \Omega_x} \frac{1}{\sqrt{2\pi}\sigma_1} \exp(-\frac{(\mu_1(\boldsymbol{x}) - I(\boldsymbol{y}))^2}{2\sigma_1^2(\boldsymbol{x})} \mathrm{d}\boldsymbol{y} \\ p_2 &= \int\limits_{\Omega_2 \cap \Omega_x} \frac{1}{\sqrt{2\pi}\sigma_2} \exp\left(-\frac{(\mu_2(\boldsymbol{x}) - I(\boldsymbol{y}))^2}{2\sigma_2^2(\boldsymbol{x})} \right) \mathrm{d}\boldsymbol{y} \end{aligned} \tag{5.8}$$

where p_1 and p_2 denote the probability density distribution of the object and background, respectively.

5.3.2 Slice-by-Slice Correction

The segmentation results obtained from the first stage are rather accurate. However, due to the complexity of medical images and associated artefacts, these may contain outliers, such as kissing non-arterial vessels and other fragments (e.g. as shown in Fig. 5.5a). In order to further improve segmentation performance, we propose a novel slice-by-slice correction scheme to the resulting images obtained in the first stage of the segmentation. We assume that the coronary arteries can be modelled as a tree structure, originating from the ostium (located in the descenting aorta at the top of the heart) and that their transaxial cross-sectional segments taken from the top to the bottom can only split but not merge over the frame sequence (see Fig. 5.6).

It can be observed from the zoomed image at the right of Fig. 5.5b that the non-arterial vessel appears darker than the coronary, and therefore, it can be distinguished from the coronary by intensity difference. However, two-phase level set methods (i.e. using one level set function for segmentation) can only separate two homogenous regions. Images with more than one object regions cannot be correctly segmented using such models. In this research, we employ a multiple regions competition-based level sets method to correct the resulting image slice-by-slice. In the proposed method, each homogenous region (object) is represented by a level set function, and thus, other non-arterial structures can be identified and removed from the segmentation. As described by Brox and Weickert [25], for a fixed number of objects to be segmented, the evolution equations of multiple regions level sets for image segmentation can be defined as:

$$\frac{\partial E(\phi_i)}{\partial t} = H'(\phi_i)\left(e_i - \max_{\substack{H'(\phi_j)>0 \\ j\neq i}}(e_j, e_i - 1)\right)$$
$$e_k = \log p_k + \tfrac{\nu}{2}\,\mathrm{div}\left(\tfrac{\nabla\phi_k}{|\nabla\phi_k|}\right) \tag{5.9}$$

Since the term log(p_1) is always negative, the evolution of a contour based on this force alone will eventually lead to the level set becoming negative everywhere, and thus eventually shrinking to a single point. The maximum operator ensures that the contour expands outwards with a constant speed, when there is no competition around the zero level set of the current embedding function. If there are multiple regions within a narrow band of current zero levels of the embedding function, the contour will evolve according to the maximal force in the interface. However, the scheme would quickly expand the current contour if only one curve is present, thus, moving the contour away from its initial location and capturing undesired boundaries. By making use of the fact that the segmentation obtained in the first stage is almost accurate, we allow the curve to evolve according to multiple-phase energy when there is no competition nearby, while, when multiple contours are presented in an interface, the points in the interface will move together subject to the strongest force across all the regions. The steps of the slice-by-slice correction algorithm are as follows:

1. Determine the first axial image which contains coronary arteries using the segmented images obtained in the first stage.
2. Assign each connected object to a level set function in the starting slice.

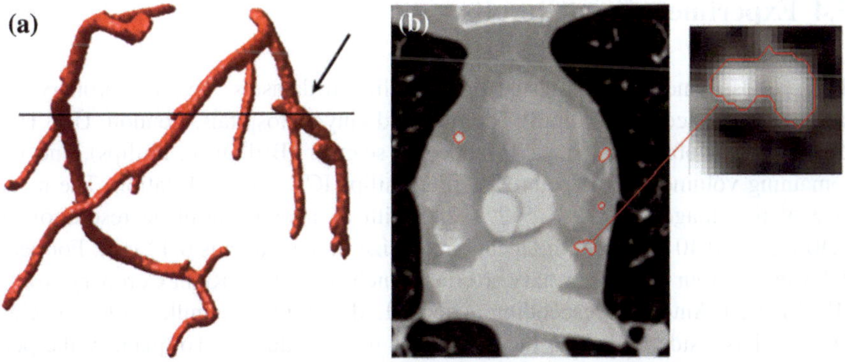

Fig. 5.5 Segmentation results obtained from the first stage of the proposed algorithm. **a** 3D surface reconstruction of the coronary arteries with a touching non-arterial vessel (shown by the *arrow*). **b** The cross-sectional image taken from the volume. The closed up image shows the intensity characteristics of the vessel in an axial image. The non-arterial vessel (*arrowed*) appears darker than the artery

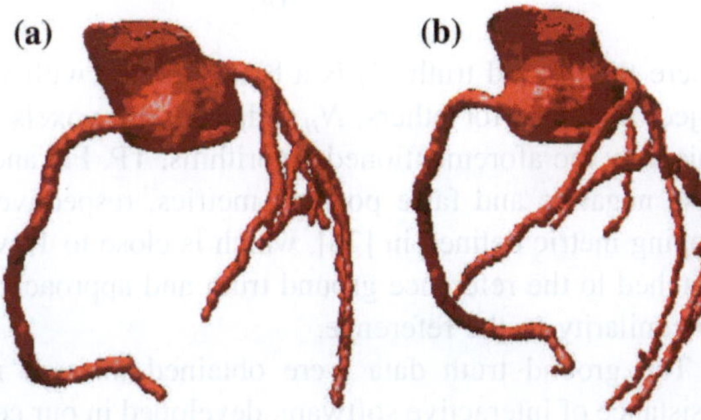

Fig. 5.6 3D reconstructed image of the coronary arteries and part of the aorta. **a** and **b** are 3D surface rendering images reconstructed from different view of points, respectively. They have shown that the arteries originate from the aorta

3. Evolve the curves in terms of the energy function defined in Eq. 5.9. To speed up convergence, the segmentation mask obtained from this slice is used as an initial estimate in the following slice.
4. Detection of non-arterial objects. We compare the initialization against the segmentation mask obtained from the first stage. If there is a connected component which does not touch any known components in the initialization mask, then the object is considered as a non-arterial component, and a level set is assigned to it.
5. If the area of the level sets is less than one pixel, we remove such level set and decrease the number of level sets functions.
6. If the current frame is not the last frame containing coronaries, go to step 3.

5.4 Experiments and Results

The proposed method was tested on real clinical datasets. Twelve coronary CT volumes were acquired from St Thomas' and Guys' Hospitals, London, UK. Two of them were obtained with a 16-slice CT scanner (Brilliance, Philips), and the remaining volumes were acquired with a Philips ICT-256 workstation. The mean size of the images is 512 × 512 × 286 with an average in-plane resolution of 0.40 mm × 0.40 mm, and the mean voxel size in the z-axis is 0.42 mm. For each CTA image, four major coronary arterial branches, namely, Right Coronary Artery (RCA), Left Anterior Descending Artery (LAD), Left Circumflex artery (LCX) and one large side branch of the coronaries were evaluated. To quantify the performance of the resulting segmentation, four metrics were used to validate the results, specifically:

$$\begin{aligned} \text{TP} &= \frac{N_B \cap N_R}{N_R}, \quad \text{FN} = \frac{N_R - N_B \cap N_R}{N_R} \\ \text{FP} &= \frac{N_B - N_B \cap N_R}{N_R}, \quad \text{OM} = 2 \cdot \frac{N_B \cap N_R}{N_B + N_R} \end{aligned} \tag{5.10}$$

where the ground truth N_R is a binary image with voxels labelled to one for the object and zero for others, N_B indicates the voxels, which are segmented as the object by the aforementioned algorithms. TP, FN and FP denote the true positive, false negative and false positive metrics, respectively. OM represents the overlapping metric defined in [26], which is close to 1, when the segmentation is well matched to the reference ground truth and approaches zero when the results have no similarity to the reference.

The ground truth data were obtained through manual delineation with the assistance of interactive software, developed in our centre. To construct the ground truth data, the CT volume was firstly resampled as a successive cross-sections, which is perpendicular to the course of the arteries, as shown in Fig. 5.7. Then, the luminal area was manually annotated by trained biomedical engineering research students (see Fig. 5.8a). The software will fit the manually delineated curve to a circle by solving the associated least squares problem (see Fig. 5.8b). The software then records the coordinates of the centre and the associated radius. In order to ensure the correct generation of the ground truth data, the luminal area of the artery is required to be specified by the observer at least every 3 mm. By doing so, the ground truth data for each major branch of the arterial tree contains on average 48 central axis points and the associated radii, which takes approximately half an hour to complete. Next, the centreline points were uniformly resampled with a distance at 0.05 mm (roughly 0.1 voxel), and the associated radii were determined via linear interpolation. To construct a closed surface of the ground truth data, we firstly generate the boundary points of the artery based on the centreline and radius information, which is depicted in Fig. 5.9a. Then, the outer surface of the artery can be reconstructed using the ball pivoting algorithm, as illustrated in Fig. 5.9b.

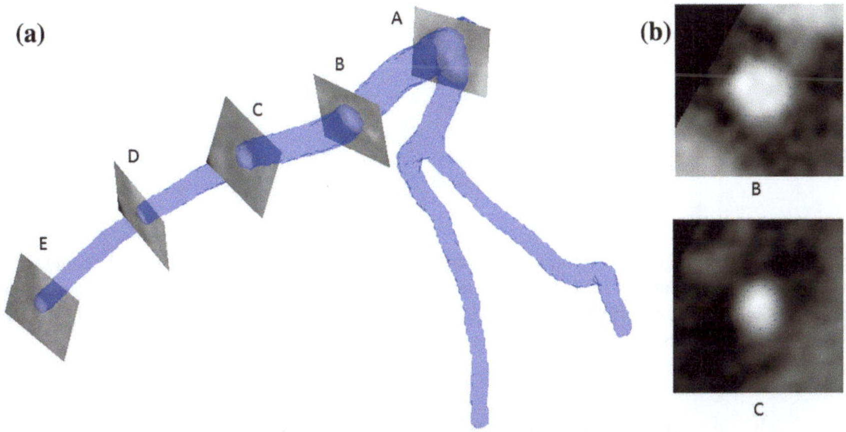

Fig. 5.7 Illustration of the resampled cross-sectional images of the CTA data for the construction of the ground truth data. **a** The segmented coronary artery is represented as 3D surface (semi-transparent structure), **b** Cross-sectional images of the artery, obtained by slicing the volume data using the planes perpendicular to the orientation of the vessel

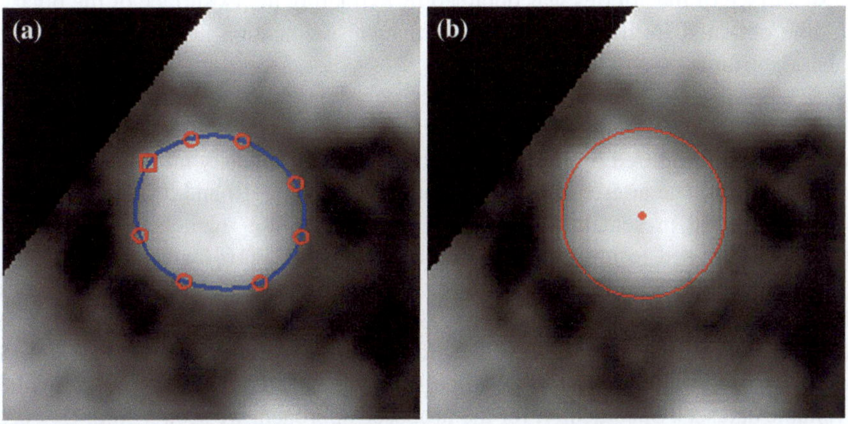

Fig. 5.8 Determination of the ground truth data based on the successive cross-sections. **a** An example of the annotation of the observer (show in *blue*), the *red circles* are the control points determined by the observer (the *square* denotes the starting control point), **b** The resulting ground truth data determined by the software in *red*. The *red dot* indicates the centre and the *circle* represents the radius of the artery at this cross section

Since the manual segmentation procedure is very time consuming, only four major branches, i.e. right coronary artery (RCA), left anterior descending artery (LAD), left circumflex artery (LCX) and one large side branch of the coronaries, were chosen for evaluation. In addition to the metrics defined in Eq. 5.10, the Hausdorff distance was also applied to measure the difference between the segmented vessel surface and the manually delineated ground truth data. The Hausdorff distance is defined as:

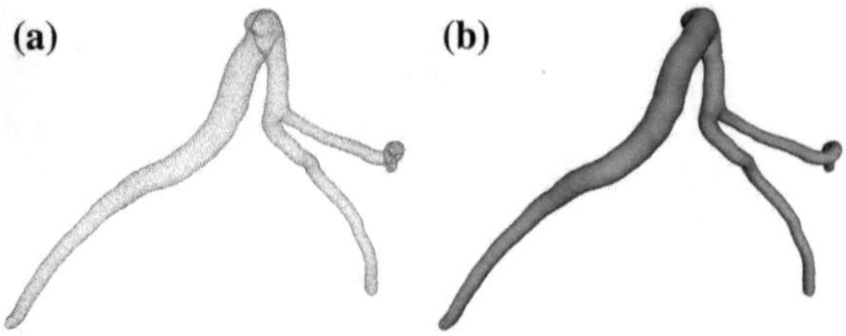

Fig. 5.9 Construction of the ground truth surface from manually delineated vessel boundaries. **a** The boundary points of the *left* coronary artery constructed using the centreline and the corresponding radius information obtained via manual annotation, **b** The outer surface of the artery reconstructed based on the boundary points shown in (**a**)

$$\mathrm{d}_H(\boldsymbol{X}, \boldsymbol{Y}) = \max\{\sup_{x \in \boldsymbol{X}} \inf_{y \in \boldsymbol{Y}} \mathrm{d}(x, y), \sup_{y \in \boldsymbol{Y}} \inf_{x \in \boldsymbol{X}} \mathrm{d}(x, y)\} \tag{5.11}$$

where ***X***, ***Y*** are the vertices of the mesh surfaces of the arteries corresponding to the segmentation results and the ground truth, respectively, and d(x, y) measures the Euclidean distance between points x and y belonging to vertices ***X*** and ***Y***. The mesh surface of the arteries was obtained by extracting the isosurface of the binary volume obtained from the segmentation/manual delineation, using the marching cube algorithm.

To demonstrate the efficiency of incorporating local intensity features into active contour functional, we compare the performance of the proposed method with the work proposed by [16], which utilises global intensity information alone, in extraction of the arteries in clinical images. In Figs. 5.10 and 5.11, Tables 5.1, 5.2 and 5.3, we present the comparison of the resulting segmentation obtained using the proposed technique and Yang et al.'s method [16], with respect to the ground truth data. The initial surface for the active contour models was obtained through the application of a Hessian-based vessel enhancement filter [27]. The tuning parameters of both of the two techniques were empirically determined from a training set, which consisted of three CT studies randomly selected from the available datasets. Specifically, for the proposed approach, we set $u = 0.2$, $v = 0.1$ and the radius of localised kernel was set to 7 voxels. The proposed approach was implemented in MATLAB (R2010b) on a standard specification PC (Dell Precision T3500, Inter(R) Xeon(R) CPU at 2.67 GHz), and the average execution time was found to be 80 s for extraction of the entire coronary trees. Yang et al.'s algorithm, on the other hand, requires roughly 45 s to carry out the same process.

As shown in Table 5.1, the mean TP rate and OM metric for the proposed method were found to be 91.1 % and 0.776, respectively, which indicate that the proposed method is able to correctly extract the major branches of the coronary arteries (see Fig. 3.10a–c). Meanwhile, the high values of the FP rate (39.2 % on average) mean that the proposed method over-segments the arteries, as illustrated in Fig. 5.11, where the segmentation results were shown on the 2D axial image as contours. In these axial images, the red contours represent the ground truth

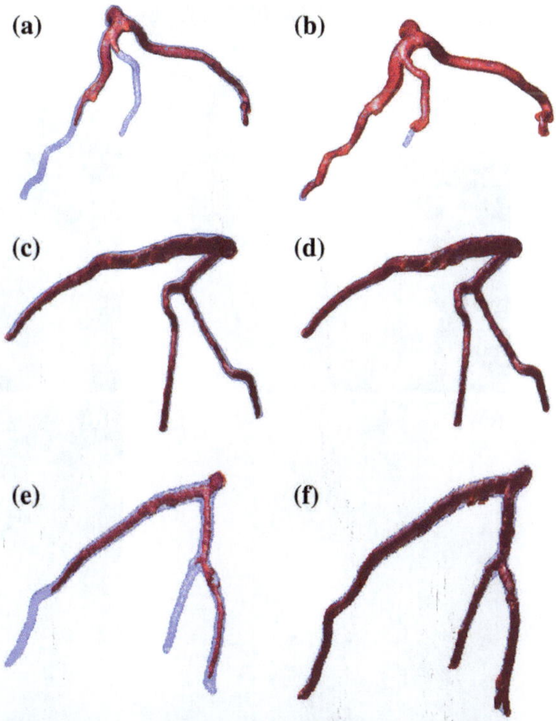

Fig. 5.10 Comparison of the resulting segmentation for datasets #3, #5 and #11 obtained using the proposed model (**a**), (**c**) and (**e**) and Yang et al. [16] technique (**b**), (**d**) and (**f**), with respect to the ground truth data. The semi-transparent surface denotes the outer surface of the ground truth data, and the resulting segmentation is depicted in *red*

boundary, and the blue and black curves represent the segmentation obtained using the proposed method and Yang et al. [16] algorithm, respectively. Since the ground truth data were modelled as circular cross-sectional tubes, this would lead to the resulting ground truth data under-estimating the true vessel surfaces. However, in terms of the voxel-wise measurements, the average value of the Hausdorff distance metric was found to be 0.73 voxels, which implies that the proposed method is capable of extracting the luminal surfaces of the arteries with sub-voxel accuracy.

In terms of the FP rate, the Yang et al. [16] method outperforms the proposed approach, with the average value found to be 16.9 %. However, we note that the TP and OM metrics of their segmentation results are significantly lower than those of the proposed technique, with the average values being 53.8 % and 0.556, respectively. In addition, by observing the statistics presented in Tables 5.2 and 5.3, we note that both the TP rate and the OM metric vary significantly, with the TP rate ranging from 23.4 to 89.9 %, while the minimum and maximum values of the OM metric were found to be 0.318 and 0.717, respectively. These observations imply that Yang et al. [16] model under-segments the coronary arteries and is only able to extract partial branches of the arterial tree, which is illustrated in Figs. 5.10b, d, e. The reason for this is that the technique, solely relying on global intensity statistics, is sensitive to image contrast and brightness changes. The method is able to correctly extract the arteries when the intensity densities are evenly distributed along the vessel. However, uneven intensity distribution is

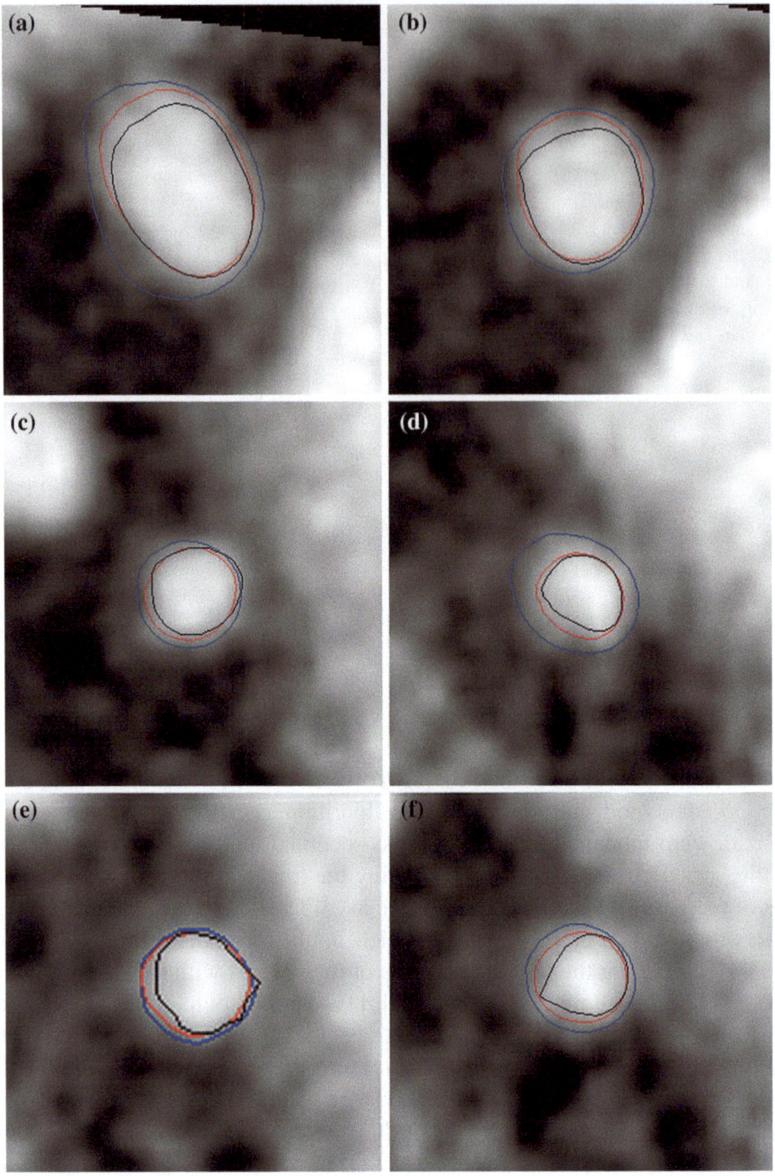

Fig. 5.11 2D transaxial images illustrating the segmentation results. **a** and **b** depict the resulting segmentation on 2D cross-sectional images randomly taken from dataset #3, **c** and **d** depict the resulting segmentation on 2D axial image from dataset #5, The examples of the segmentation results on data #11 are illustrated in (**e**) and (**f**). The *red* contour represents the manually delineated ground truth, while the segmentation obtained from the proposed method and Yang et al. [16] method are shown in *blue* and *black* contours, respectively. Note that, the cross-sectional images were up-sampled by a factor of five using linear interpolation to increase the resolution

Table 5.1 Comparison of the 3D CTA segmentation results between the proposed method and Yang et al. [16] technique (overall)

Rate	Methods	
	Proposed method mean	Yang et al. [16] method mean
TP (%)	91.1	53.8
FP (%)	39.2	16.9
OM	0.776	0.556
Mean (d_H)	0.730	1.07

Table 5.2 Comparison of the 3D CTA segmentation results between the proposed method and Yang et al. [16] technique: datasets #1–6

Metrics	Methods	3D CTA images					
		1	2	3	4	5	6
TP (%)	Proposed	**94.1**	**93.9**	**93.1**	**92.8**	**97.0**	**93.2**
	Yang et al.	47.6	50.3	52.1	25.6	89.9	87.9
FN (%)	Proposed	**32.4**	**28.4**	**43.1**	**38.3**	**45.5**	**29.4**
	Yang et al.	1.31	8.39	13.6	3.16	60.7	68.9
FP (%)	Proposed	**0.831**	**0.731**	**0.791**	**0.803**	**0.800**	**0.837**
	Yang et al.	0.639	0.634	0.629	0.397	0.717	0.685
OM	Proposed	**0.623**	**1.02**	**0.670**	**0.833**	**0.623**	**0.782**
	Yang et al.	0.865	1.35	1.00	1.09	0.767	0.891
Mean(d_H)	Proposed	**94.1**	**93.9**	**93.1**	**92.8**	**97.0**	**93.2**
	Yang et al.	47.6	50.3	52.1	25.6	89.9	87.9

Table 5.3 Comparison of the 3D CTA segmentation results between the proposed method and Yang et al. [16] technique: datasets #7–12

Metrics	Methods	3D CTA images					
		7	8	9	10	11	12
TP (%)	Proposed	**90.1**	**89.0**	**95.3**	**80.5**	**86.5**	**87.8**
	Yang et al.	51.2	49.2	88.4	53.6	23.4	26.7
FN (%)	Proposed	**41.6**	**38.6**	**51.8**	**35.2**	**42.2**	**44.7**
	Yang et al.	3.16	1.57	16.3	10.0	9.55	6.18
FP (%)	Proposed	**0.778**	**0.782**	**0.713**	**0.744**	**0.756**	**0.755**
	Yang et al.	0.663	0.318	0.580	0.655	0.353	0.402
OM	Proposed	**0.759**	**0.719**	**0.620**	**0.561**	**0.769**	**0.783**
	Yang et al.	1.07	1.63	0.976	1.15	1.24	0.861
Mean(d_H)	Proposed	**90.1**	**89.0**	**95.3**	**80.5**	**86.5**	**87.8**
	Yang et al.,	51.2	49.2	88.4	53.6	23.4	26.7

commonly encountered in coronary CTA, because of the concentration attenuation of the contrast agent and acquisition noise. In this case, the Yang et al. [16] approach can only extract the proximal segments of the arteries, since distal

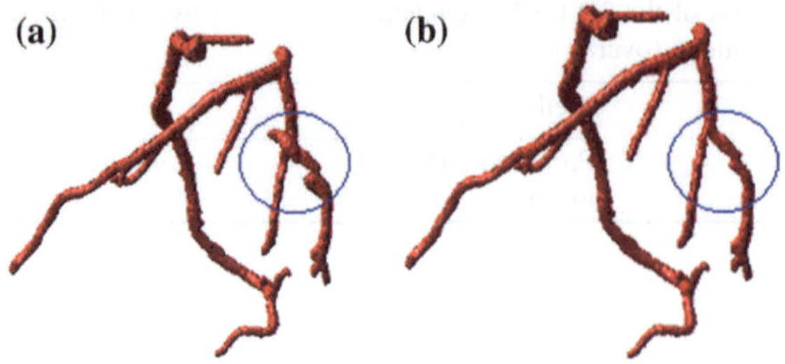

Fig. 5.12 The comparison **a** before and **b** after the correction process (datasets #7). The touching non-arterial vessel (within the *blue circle*) has been removed

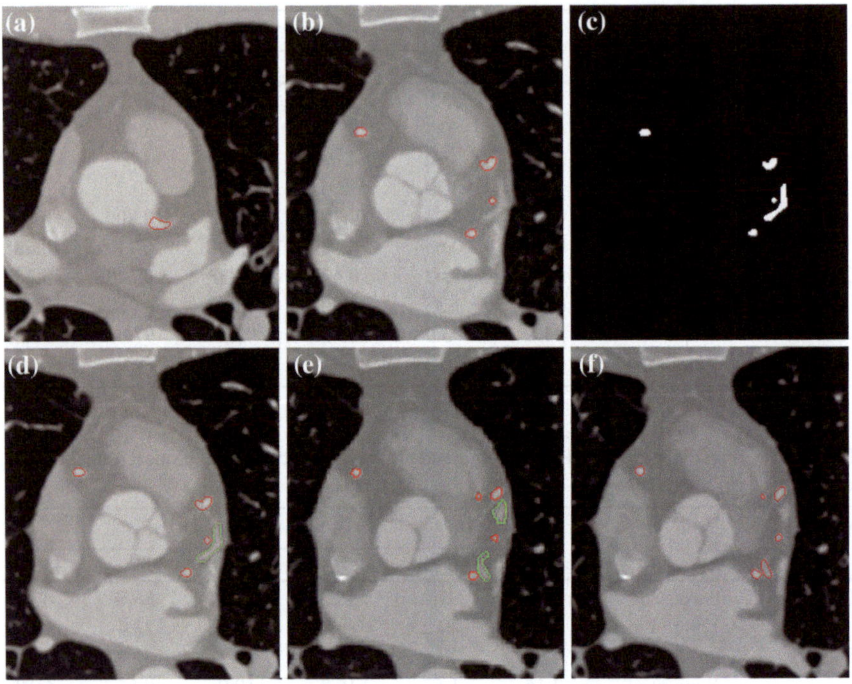

Fig. 5.13 Transaxial slices illustrate the slice-by-slice correction algorithm. **a** The first slice contains the artery (delineated in *red*), **b** The artery is tracked through slices, **c** The binary image obtained from the first stage segmentation, this is the first slice that contains the kissing vessel (*arrowed*), **d** and **e** Kissing vessel is identified and tracked over slices (depicted in *green* contours), **f** The original segmentation in same transaxial slice as shown in (**e**)

segments have relatively lower intensity values and lack image contrast. In addition, we also found that the TP and OM rates of the proposed method tend to decrease when approaching the distal and small segments of the arteries. However, clinically significant coronary lesions are usually identified in the main and proximal branches of the arteries, which can be well defined by the proposed

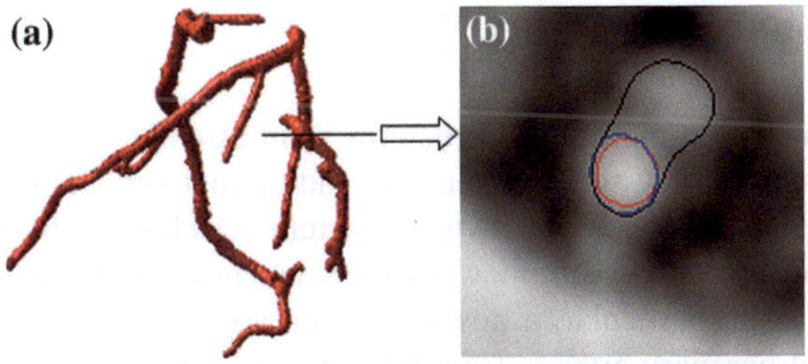

Fig. 5.14 The efficiency of the proposed slice-by-slice correction algorithm demonstrated by cross-sectional view image. **a** 3D volume data and **b** Cross-sectional view image randomly taken from the 3D volume data, illustrating the resulting segmentations with and without the correction step are shown in *blue* and *black*, respectively. The *red* contour represents the reference boundaries of the vessel obtained through manual delineation

Table 5.4 Comparison of the segmentation results before and after the application of the correction scheme

Rate	Before correction	After correction
TP (%)	82.4	85.3
FN (%)	17.6	14.7
FP (%)	294	22.8
OM	0.380	0.820

method. Nevertheless, we can conclude that our technique is able to delineate the vessel boundaries in clinically important coronary segments with a level of variability similar to those obtained through manual segmentation.

Two CTA studies were affected by the presence of kissing vessels (i.e. datasets #7 and #11). The slice-by-slice correction scheme was subsequently applied after the first step of the segmentation. Figure 5.12 illustrates the 3D surface reconstruction image before and after applying the slice-by-slice correction algorithm. Figure 5.13a shows the first slice, which contains the cross-sectional segments of the coronary arteries. The false positives, i.e. the kissing vessel artefacts, were identified and then tracked in Fig. 5.13b–e. The boundaries of the coronary arteries are shown in red, and the green curve represents the kissing vessel structures. In Fig. 5.13f, we illustrate the initial segmentation of the axial image shown in Fig. 5.13e prior to the correction.

Instead of using the entire volume to quantify the performance of the slice-by-slice correction, we conduct the assessment on a 2D axial image basis. The resulting segmentation with and without the slice-by-slice correction is illustrated in Fig. 5.14, where the boundaries of the artery before and after correction are depicted in black and blue, respectively, and the manual segmentation is shown in red. The four metrics, defined in Eq. 5.10, were used to quantify the performance and the results are shown in Table 5.4. It can be observed from the table that the FP rate, which is primarily caused by the kissing vessels, is dramatically reduced after the application of the slice-by-slice correction algorithm (reduced from 294 to 22.8 % for arterial segments containing kissing vessels, e.g. the vessel segment shown in Fig. 5.12a, within the circle).

5.5 Conclusions and Future Work

Accurate segmentation of vascular structures is an essential and fundamental step for various clinical tasks, such as stenosis grading and surgical planning. In this chapter, we presented a computer vision system which contributes towards the development of the proposed framework for segmentation of coronary arteries in 3D CTA images. In particular, a novel two-step algorithm was proposed to efficiently segment coronary arteries from CT images by making use of both global and local intensity statistics. The global energy was designed as a normalised CDF based on the histogram of the input image data, which adapts its contribution to the overall active contour energy by considering the spatially varying properties of the artery. The kissing vessels were identified and tracked throughout the axial slices in the second stage of the segmentation based on a multi-region competition algorithm. Experimental results show that the proposed approach is able to correctly segment the major branches of the arterial tree, with an average voxel-wise distance of 0.73 voxels to the manually delineated ground truth. Furthermore, in the presence of kissing vessel artefacts, the overall performance of the segmentation can be significantly improved by the slice-by-slice correction scheme. The FP rate from these cross-sections containing kissing vessels was reduced from 294 to 22.8 %.

In terms of future research, we intend to introduce spatial/intensity correlations between frames to the correction process, since the information extracted from previous frames can be used to assist in segmentation of coronary arteries in the subsequent frame. Based on the 3D surface reconstructed from the segmentation, the geometric parameters along each of the main branches of the coronary arteries can be precisely computed, which are very useful for diagnostic purposes, such as stenosis detection and grading.

Acknowledgments The authors would like to acknowledge the support of City University, which enabled this research through the award of a University Research Studentship and Dr Gerry Carr-White and Rebecca Preston at St Thomas and Guys Hospitals for their invaluable advice and the provision of the CTA datasets.

References

1. WHO. Coronary Heart Disease (2010)
2. Netter, F.H.: Atlas of Human Anatomy, 2nd edn. Rittenhouse Book Distributors Inc, King of Prussia (1997)
3. Abe, T., Hirohata, M., Tanaka, N., et al.: Clinical benefits of totational 3D angiography in endovascular treatment of ruptured cerbral aneurysm. Am. J. Neuroradiol. **23**, 686–688 (2002)
4. Tillich, M., Hill, B.B., Paik, D.S., et al.: Prediction of aorotoiliac stent graft length: comparison of measurement methods. Radiology **220**(2), 475–483 (2001)
5. Philips. Brilliance iCT scanner. http://www.healthcare.philips.com/us/products/ct/products/ct_brilliance_ict/index.wpd (2011)

6. Chan, T.F., Vese, L.A.: Active contours without edges. IEEE Trans. Image Process **10**(2), 266–277 (2001)
7. Kompatsiaris, I., Tzovaras, D., Koutkias, V., et al.: Deformable boundary detection of stents in angiographic images. IEEE Trans. Med. Imaging **19**(6), 656–662 (2000)
8. Pieta, L., Tomczyk, A., Szczepaniak, P.S.: Medical image analysis using potential active contours. Inf. Technol. Biomed. Adv. Soft Comput. **47**, 66–73 (2008)
9. Ray, N., Acton, S.T., Altes, T.: Merging parametric active contours within homogeneous image regions for MRI based lung segmentation. IEEE Trans. Med. Imaging **22**(2), 189–199 (2003)
10. Sum, K.W., Chueng, Y.S.: Vessel extraction under nonuniform illumination: a level set approach. IEEE Trans. Biomed. Eng. **55**(1), 359–361 (2008)
11. Yezzi, A., Kichenassamy, S., Kumar, A., et al.: A geometric snake model for segmentation of medical imagery. IEEE Trans. Med. Imaging **16**(2), 199–209 (1997)
12. Yezzi, A., Tsai, A., Willsky, A.: A statistical approach to snakes for bimodal and trimodal imagery. In: The Proceedings of the Seventh IEEE International Conference on Computer Vision, Kerkyra (1999)
13. Osher, S., Fediww, R.: Level Set Methods and Dynamic Implicit Surface. Springer, New York (1999)
14. Caselles, V., Kimmel, R., Sapiro, G.: Geodesic active contours. Int. J. Comput. Vision **22**(1), 61–79 (1997)
15. Kichenassamy, S., Kumar, A., Olver, P., et al.: Gradient flows and geometric active contour models. In: Proceedings of Fifth International Conference on Computer Vision, Boston (1995)
16. Yang, Y., Tannenbaum, A., Giddens, D., et al.: Automatic segmentation of coronary arteries using bayesian driven implicit surfaces. In: Proceedings of 4th IEEE International Symposium on Biomedical Imaging (2007)
17. Nain, D., Yezzi, A., Turk, G.: Vessel segmentation using a shape driven flow. In: Proceedings of Medical Image Computing and Computer-Assisted Intervention—MICCAI 2004 (2004)
18. Chan, T.F., Vese, L.A.: A multiphase level set framework for image segmentation using the Mumford and Shah model. Int. J. Comput. Vision **50**(3), 271–293 (2002)
19. Kim, J., Fisher, J., Yezzi, A., et al.: A non-parametric statistical methods from image segmentation using information theory and curve evolution. IEEE Trans. Image Process. **14**(10), 1486–1502 (2005)
20. Lankton, S., Tannenbaum, A.: Localizing region-based active contours. IEEE Trans. Image Process. **17**(11), 2029–2039 (2008)
21. Li, C., Kao, C.Y., Gore, J.C., et al.: Minimization of region scalable fitting energy for image segmentation. IEEE Trans. Image Process. **17**(10), 1940–1949 (2008)
22. Wang, L., He, L., Mishra, A., et al.: Active contours driven by local gaussian distribution fitting energy. Signal Process. **89**, 2435–2447 (2009)
23. Tomandl, B.F., Hastreiter, P., Eberhardt, K.E., et al.: The kissing vessel-artifact: a problem occurring in the visualization of intracranial aneurysms using volume rendering and virtual endoscopy. J. Radiol. **213**, 311–314 (1999)
24. Wong, W.C.K., Chung, A.C.S.: Probabilistic vessel axis tracing and its application on vessel segmentation with stream surface and minimum cost paths. Med. Image Anal. **11**, 567–587 (2007)
25. Brox, T., Weickert, J.: Level set segmentation with multiple regions. IEEE Trans. Image Process. **15**(10), 3213–3218 (2006)
26. Zijdenbos, A., Dawant, B., Margolin, R., et al.: Morphometric analysis of white matter lesions in MR images: methods and validation. IEEE Trans. Med. Imaging **13**(4), 716–724 (1994)
27. Frangi, A.F., Niessen, W., Hoogeveen, R.M., et al.: Model based quatitiation of 3-D magnetic resonance angiographic images. IEEE Trans. Med. Imaging **18**(10), 946–956 (1999)

Chapter 6
Wavelet Compression/Reconstruction and Visualization of Pulmonary X-Ray Images for Achieving of Asbestosis Infected Patients Data

Ivica Kuzmanić, Mirjana Vujović, Slobodan Marko Beroš and Igor Vujović

Abstract An algorithm for reliable wavelet compression/reconstruction and visualization of pulmonary X-ray is presented in this chapter. Pulmonary X-rays are obtained by real patients from an asbestos factory. The aim is to make job easier to occupational medicine specialists and radiologists. Algorithm is primarily concerned for correct compression of the images to save space (digital memory space as well as space for storing X-ray films). Specialists must, according to law, save all X-ray images over 40 years. Instead of archiving X-ray films this algorithm allows saving of wavelet coefficients vectors on magnetic or optical storage. Independent radiologists confirmed that medical data is unchanged. Secondary concern is to emphasize possible asbestos-infected areas, which covers for visualization part of the work. Benefits are in monitoring of health condition, prevention of disease, early diagnostics, more reliable diagnostics, and saving space for achieving medical data.

Keywords Wavelet image compression · Preservation of medical data · Comparison of wavelet families

I. Kuzmanić (✉) · I. Vujović
Faculty of Maritime Studies, University of Split, Zrinjsko-Frankopanska 38, Split, Croatia
e-mail: ikuzman@pfst.hr

I. Vujović
e-mail: ivujovic@pfst.hr

M. Vujović
Occupational Medicine Private Practice, Trg kralja Tomislava bb, 20340 Ploče, Croatia
e-mail: mvujovic@globalnet.hr

S. M. Beroš
Faculty of Electrical Engineering, Mechanical Engineering and Naval Architecture, University of Split, Ruđera Boškovića bb, 21000 Split, Croatia
e-mail: sberos@fesb.hr

F. R. Leta (ed.), *Visual Computing*, Augmented Vision and Reality 4,
DOI: 10.1007/978-3-642-55131-4_6,

List of Abbreviations

DICOM	The Digital Imaging and Communications in Medicine
IEEE	Institute of Electrical and Electronic Engineers
JPEG	Joint Photographic Experts Group—file format
WT	Wavelet transform
DWT	Discrete wavelet transform
2D-DWT	Two-dimensional discrete wavelet transform

6.1 Introduction

Archiving medical data is demanding operation. It is very interesting in data with medical images (X-rays, mammography, tomography, etc.).

Asbestos particles cause irritation of lung tissue and reduce breathing capacity of the patients. The set of symptoms including mentioned above are called asbestosis. Furthermore, presence of asbestos increases the possibility of lung cancer.

Due to asbestos toxicity, several attempts to engineered diagnostics, visualization, and archiving have been made. According to the IEEE Xplore, the most relevant are described next.

Usage of image processing for asbestos detection starts before 1978 [1] by automating counting of asbestos fibers in air. Digitalized image was used. Inclusion of toxicity in engineering education was proposed with growth of awareness about asbestos and other dangerous materials [2]. Remote sensing with image differencing method was used to assess the condition of vegetation contaminated with asbestos [3]. X-Ray diffraction and scanning electron microscopy were performed to detect the presence of asbestos minerals in environment [4]. Pictures taken by phase contrast microscopy were used for image processing of particle detection for asbestos qualitative analysis (particle counting by using color variance of background) [5]. Automated microscopic imaging system for supporting asbestos qualitative analysis was proposed in [6], but it does not use X-rays too. Multispectral Infrared Visible Imaging Spectrometer imagery was proposed in [7] to identify roads' asphalt pavements and the asbestos cement roofs.

It has to be noted that any of these references considered humans, but rather environmental pollution. With the advancement of telemedicine, transfer of compressed medical data actualized image processing and analysis of pulmonary X-rays (from sources: analog film or digitalized X-ray devices). Telemedical applications were considered in [8]. Computer-aided diagnostics of asbestosis was proposed in [9, 10]. Telediagnostics was proposed in [11] based on [8–10]. Compression part was explained and researched in [12]. Asbestos relation to cancer was proved in [13–15], which makes diagnostics of asbestosis and asbestos-related diseases very important.

Wavelets in biomedical applications were covered in [16]. Considerable efforts were performed from 1990s till this day to apply wavelets for radiological imaging. Fidelity of reconstructed radiographic images was discussed in [17], where wavelet transform and JPEG coding were used. Wavelets were introduced in DICOM standard for image compression, because of fidelity [17, 18]. Fidelity for different wavelets and compression ratios (CR) was checked in [12].

This chapter deals with wavelet compression of the pulmonary X-rays with important condition—reconstructed images must provide the original medical information. Furthermore, suspicious areas in the reconstructed images should be emphasized to make easier diagnostics.

Benefits are in monitoring of health condition, prevention of disease, early diagnostics, more reliable diagnostics, and saving space for achieving medical data. Therefore, contribution to saving and/or prolonging human lives is the most important benefit of the proposed algorithm.

The chapter is organized as follows. In the following sections, theoretical background, proposed algorithm, results, and conclusions will be presented, respectively.

6.2 Theoretical Background

Image compression, is one of the most outstanding applications of wavelets [16, 19, 20]. Powerful compression possibilities of wavelets have been exploited in many applications, off and online, for single images and for image sequences [21]. Wavelets are incorporated in JPEG-2000 standard as well and security [22–24]. Their ability in denoising and compression often depend on thresholding. In the proposed algorithm (in the next heading), thresholding is avoided.

The basic idea of an integral representation is to describe a signal $x(t)$, that is integrable in Lebesque sense and closed on $L_2(R)$, via its density $X(s)$ with respect to arbitrary kernel $\varphi(\mathrm{t,s})$:

$$x(t) = \int\limits_S X(s)\varphi(t,s)\,\mathrm{d}s \quad t \in T \subseteq L_2(R) \tag{6.1}$$

The wavelet transform $W(a,b)$ of a continuous-time signal $x(t)$ is defined as:

$$W(a,b) = |b|^{\frac{-1}{2}} \cdot \int\limits_{-\infty}^{+\infty} x(t) \cdot \psi^*(\frac{t-a}{b}) \cdot \mathrm{d}t \tag{6.2}$$

where b is scaling parameter, a translation parameter, and $\psi(t)$ wavelet. Thus, the wavelet transform can be computed as the inner product of $x(t)$ and translated and scaled versions of the wavelet. If $\psi(t)$ is considered to be a bandpass impulse response, then the wavelet analysis can be understood as a bandpass analysis.

Time and frequency resolution of WT depends of b. For high analysis frequencies, good time localization but poor frequency resolution can be achieved. When using a transform in order to get better insight into the properties of a signal, it should be ensured that the signal can be perfectly reconstructed from its representation. Otherwise the representation may be completely or partly meaningless.

Images are 2D functions of intensity. The analysis starts with defining a two-dimensional scaling and wavelet functions:

$$s_{\phi\phi}(x,y) = \phi(x)\cdot\phi(y) \tag{6.3}$$

$$s_{\psi\psi}(x,y) = \psi(x)\cdot\psi(y) \tag{6.4}$$

If $f_0(x,y)$ is the projection of $f(x,y)$ on the space V_0 generated by $s_{\phi\phi}(x,y)$, we get:

$$\begin{aligned} f_o(x,y) &= \sum_{i=-\infty}^{\infty}\sum_{j=-\infty}^{\infty} a_o(i,j)\cdot s_{\phi\phi}(x-i,y-j) \\ a_o(i,j) &= <f(x,y),\quad s_{\phi\phi}(x-i,y-j)> \end{aligned} \tag{6.5}$$

The result is four sets of coefficients: approximation and horizontal, vertical and diagonal direction details. During years, several transforms (curvelets, conturelets, edgelets, etc.) emerged from wavelets [25], which define details in different manners, but these are not concern of the chapter. These transforms can be used when rotation is likely to happen. However, X-rays are always taken in fixed position. Therefore, there is no reason to introduce other transforms. Anyway, algorithm is in general enough to be modified for these transforms as well.

Image compression is usually obtained by thresholding wavelet coefficients. Some efforts were made for compression by downsampling of coefficients vector. In such case, interpolation is used in reconstruction.

6.3 Proposed Algorithm

Algorithm begins with the compression part. This part enables archiving of X-rays in vector of wavelet coefficients. When doctor wants to see the image, reconstruction part is used. After reconstruction, visualization is performed. Diagnostics is facilitated by diagnostics visualization part of the algorithm by emphasizing suspicious areas with different color. If color does not appear on the screen, this means that patient is not candidate for asbestosis (there is no suspicious shadows in the lung image).

Proposed algorithm can be explained in several steps (shown as algorithm in Fig. 6.1). Decomposition part has 6 steps as follows.

Step 1: Wavelet decomposition of the pulmonary X-ray image.
2D DWT is implemented with Matlab command wavedec2, which includes packaging the data into vector:

Fig. 6.1 Compression part of the algorithm

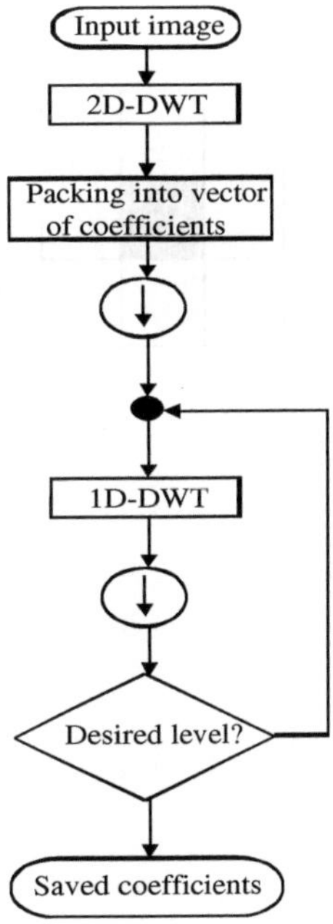

$$\mathbf{C} = [A(N)|H(N)|V(N)|D(N)|\ldots \\ H(N-1)|V(N-1)|D(N-1)|\ldots \\ \ldots \\ H(1)|V(1)|D(1)] \tag{6.6}$$

where A denotes approximation, H horizontal details, V vertical details, and D diagonal details.

Step 2: Downsampling of the vector C.
In this step, both approximation and details are downsampled to reduce data.

Step 3: 1D DWT of the vector C.
Since produce algorithm is 1D vector, new vector formed will be:

$$C_i = [A_i(N)|H_i(N)|V_i(N)|D_i(N)|H_i(N-1)| \\ V_i(N-1)|D_i(N-1)|\ldots H_i(1)|V_i(1)|D_i(1)] \tag{6.7}$$

Step 4: Downsampling of the vector C_1. Downsampling is organized to reduce number of horizontal coefficients at first. Then vertical and diagonal coefficients are downsampled, respectively. Result is new vector:

$$C_i' = [A_i(N)|H_i(N)|V_i(N)|D_i(N)|H_i(N-2)| \\ V_i(N-2)|D_i(N-2)|\ldots H_i(j)|V_i(j)|D_i(j)] \tag{6.8}$$

where j can be 1 or 2 depending on parity (weather N is even or odd).

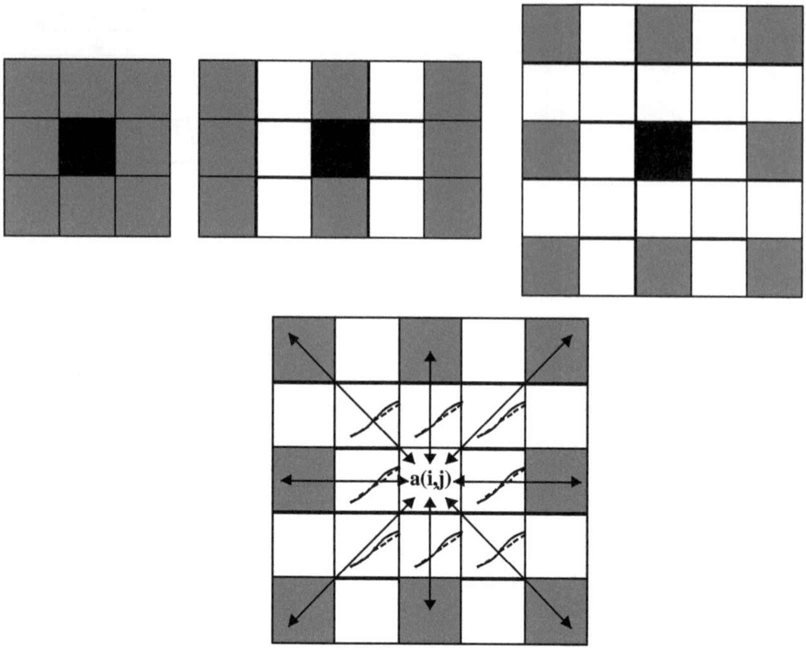

Fig. 6.2 Reconstruction process in the 2D step

Step 5: Repetition of the Steps 3 and 4 until desired level is reached. In the research, experiments were made to find maximum number of repetition for standard wavelets.

Step 6: Output of the algorithm is the coefficients' vector. The vector is saved in storage. Media for saving data can be magnetic or optical.

Two questions are of importance when the vector is storage. The first one is medical value of the stored data. The second one is how to use it. As stated previously, medical experts confirmed medical value of data obtained by algorithm. But, to be able to see the images, first images have to be reconstructed from coefficients' vector. Medical experts cannot evaluate coefficients, but images. So, the second part of the proposed system is the algorithm for visualization of the compressed data.

Algorithm for visualization of the compressed data involves reconstruction and interpolation of coefficients. The proposed reconstruction procedure is illustrated in Fig. 6.2. The proposed algorithm is shown in Fig. 6.3.

Step 1. Input to the algorithm is vector of coefficients stored in previous phase. This phase is activated when request for visualization occur.

Step 2. Input vector is upsampled by introducing zeros. This is performed columnwise, which means that number of columns is approximately doubled.

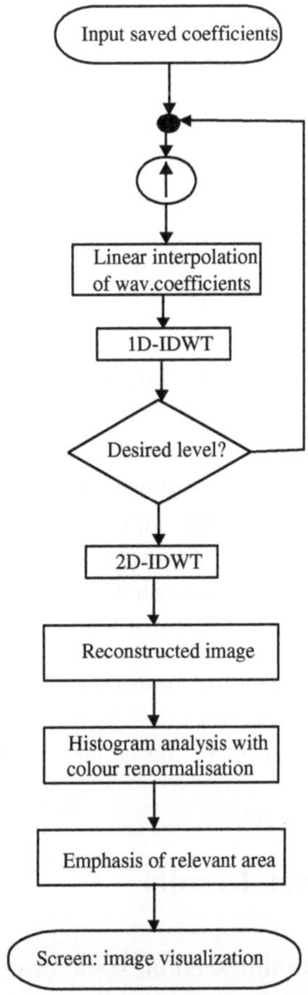

Fig. 6.3 Algorithm for visualization of the reconstructed data

Step 3. Nonlinear interpolation of coefficients at places of zeros is performed. B-spline wavelets were not used, but rather spline interpolation of wavelet coefficients in the vector. This step takes the most processor time if programming technique is weak. Recently, several programming techniques were developed to accelerate these operations. Therefore, older versions of Matlab operate slower than newest.

Step 4. Since vector is 1D type of data, 1D IDWT is performed. Notice that reconstruction cannot be always performed without resizing vector introducing zero at the end or deleting last zero.

Step 5. Repetition of Steps 2 to 4 while desired level is reached.

Step 6. 2D-IDWT is performed.

Step 7. Reconstructed image is obtained, which can be used to compare with original. If the diagnoses from reconstructed image and from the original are the same, than algorithm has medical value. Result must be 100 % unchanged if the original can be destroyed. In that case, saving of storage space can be achieved.

Step 8. In this step, shadows are detected and emphasize with pointing color to make diagnostics easier. Shadows are detected by histogram method. In the method, suspicious pixels are detected by colors of asbestos shadows. It is easy said that suspicious colors will be painted differently. However, histograms depend on many factors, which are not necessarily the same. Therefore, suspicious colors can be different on different X-ray machines. Even if the machinery is the same, if there is no direct digital transfer of image, scanning process can be different and produce different histogram. That is the reason why automatic detection of asbestos shadows should be avoided. It is better that human picks suspicious color and then algorithm emphasizes chosen color in the image of the lung. Based on the research on, approximately, 200 real patients, we determined critical colors. However, it does not mean for sure that it works everywhere. This is the reason to choose manual selection of colors. With this selection, color renormalization is performed and new adopted histogram is obtained.

Implementation of the wavelet transform depends on the programmer's choice: it can be classic filter implementation or lifting. Lifting algorithm is faster and the results are the same as corresponding filters.

6.4 Results

In this section, some of results are presented in area of compression and medical fidelity.

Since there were a lot of discussion in scientific community about image format, which can be allowed for medical imaging, it can be concluded that the safest way is to use formats which does not change original data in any way (i.e., bitmap maps every pixel). If original data is processed in some sort of coding (maybe with some threshold), risk of losing data exists, because some medically important detail can be lost.

Except image format, other questions came up as well, i.e., gray image versus color or which resolution is satisfactory. According to DICOM (eng. The Digital Imaging and Communications in Medicine) standard, minimal resolution for medical images should be 300 dpi [26]. Except bitmaps and tiff format, JPEG was tested in the research. JPEG images were compressed to size of approximately 350 KB. Original images take around 4.5 MB. Therefore, compression ratio (CR) for analyzed images in JPEG format is:

Table 6.1 Absolute size of output files to be saved

Level	Min. size of saved file	Type of wavelet	Max. size of saved file	Type of wavelet
1	17.253 MB	Haar	18.9 MB	Sym2
2	8.627 MB	Haar	9.457 MB	Sym2
4	2.158 MB	Haar	2.365 MB	Sym2
8	540 KB	Haar	592 KB	Sym2
10	35 KB	Haar	38 KB	Sym2
14	3.9 KB	Haar	4 KB	Rbio1.3

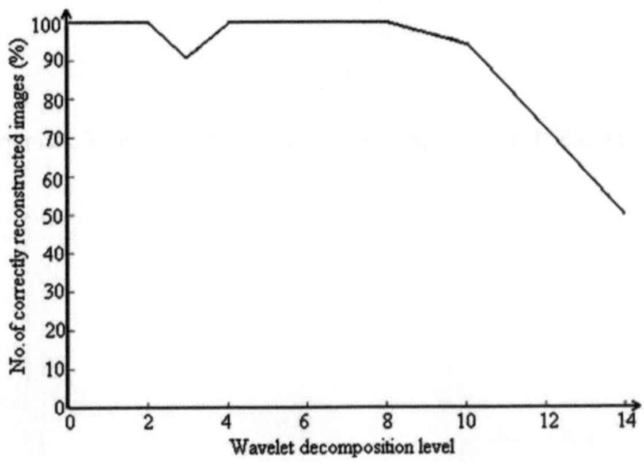

Fig. 6.4 Medical fidelity of the reconstructed images

$$CR = 4.5 * 1048576/1024 * 350 = 13.6 \quad (6.9)$$

Results obtained by proposed algorithm are presented in Table 6.1. Haar wavelet has minimum moments and simplest description. This is the reason for minimum size of the output file. Problem of Haar is granular structure of the reconstructed image.

Figure 6.4 shows fidelity of the reconstructed images versus decomposition level. Biorthogonal, reverse biorthogonal, and Symlet wavelets do not enter the error before ninth level of the decomposition. However, Daubechies standard wavelets do enter error at the third level. That is the reason for excluding it from the further consideration on the highest levels. Therefore, error is reduced. It is imperative to understand that 0 % error does not mean that reconstructed image is unchanged, but that medical diagnosis is unchanged. An example of reconstructed image is in Fig. 6.5.

An example of what to expect is in Fig. 6.6, where original and reconstructed images are compared. Reconstructed images are a bit smoother and blur is introduced, but important details are still observable.

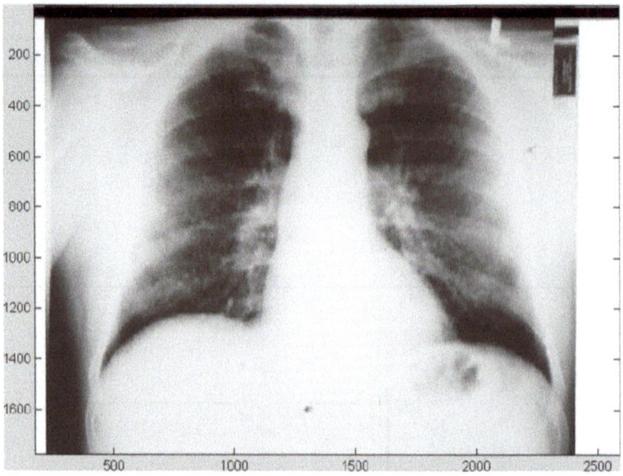

Fig. 6.5 Reconstructed image from the 14th level of decomposition

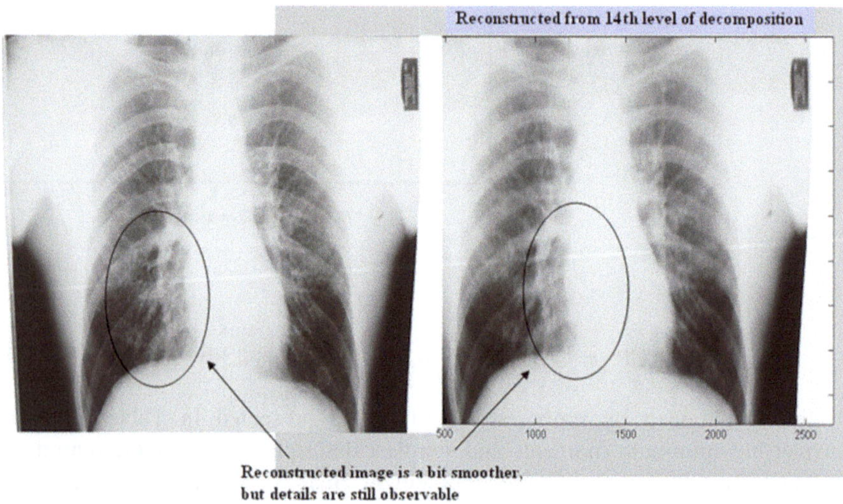

Fig. 6.6 Comparison of the original and reconstructed image

Figures 6.7, 6.8, 6.9, 6.10 are the final part of the proposed algorithm—part of visualization. Figure 6.7 is ROI designated by operator. Figures 6.8, 6.9, 6.10 are different options in visualization chosen by operator.

An example of the X-ray processed with commercial software (CR = 100) is shown in Fig. 6.11.

Comparison with the commercial software:

- Commercial software uses decompositions till 10th level. Proposed algorithm is tested till 14th level of decomposition.
- Commercial software does not use standard wavelets like in the proposed algorithm.

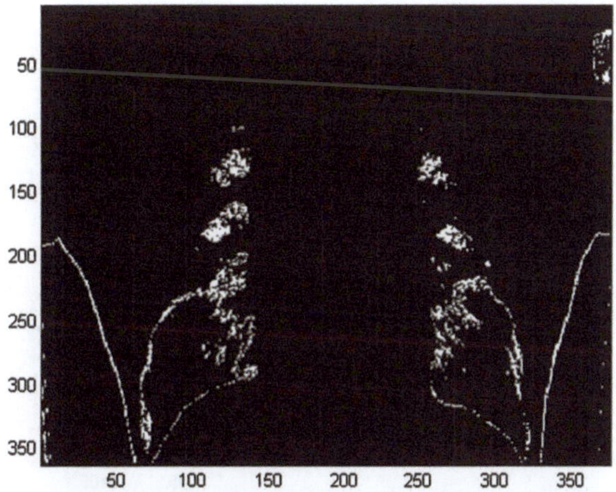

Fig. 6.7 Selected regions for visualization

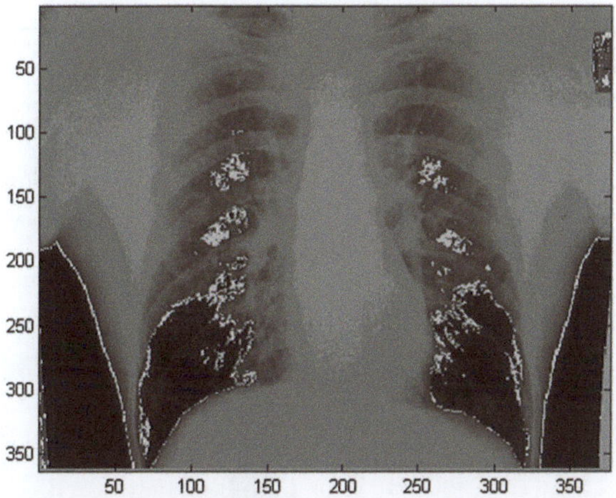

Fig. 6.8 Selected areas with defined color

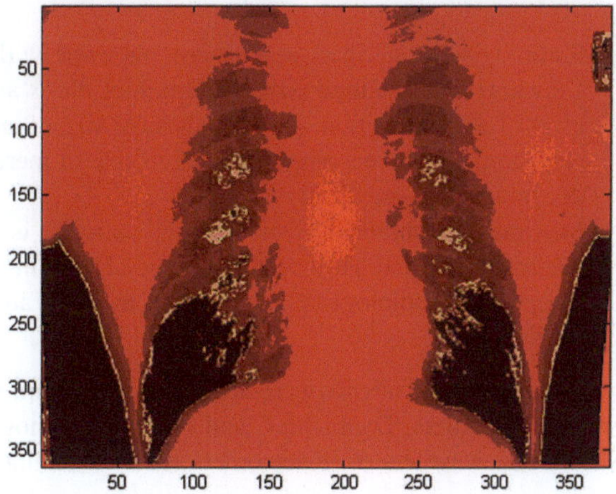

Fig. 6.9 Emphasized desired tones with wanted color

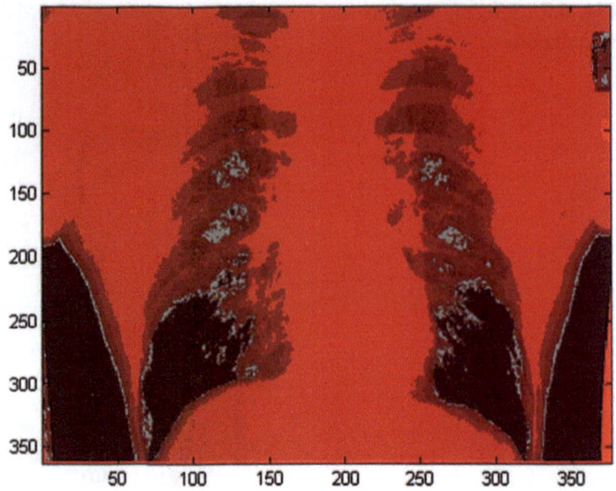

Fig. 6.10 Emphasized desired tones with different color

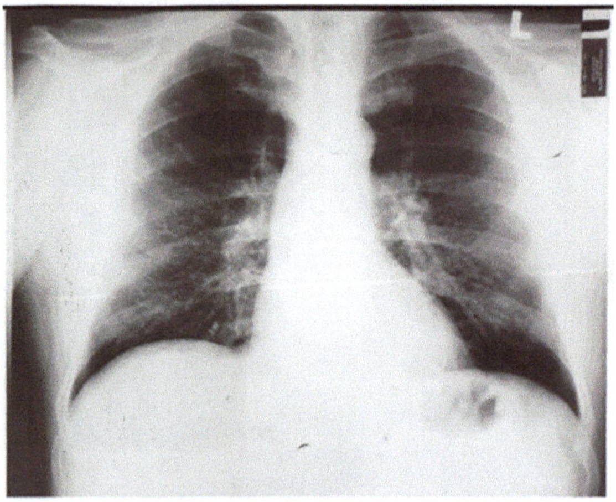

Fig. 6.11 Image compressed and reconstructed by commercial software [27]

- Radiologically verified results are obtained with CR = 24 for commercial software. The obtained size of the output file is around 190 KB. For the same level of decomposition, output files have size of around 37 KB in the proposed algorithm. However, output files have size of merely 4 KB at the 14th level of decomposition (medically verified).
- Maximal CR obtained by commercial software is 400, but such images are not medically verified. However, proposed algorithm is verified medically by three independent radiologists, as requested by ILO [28]. Obtained CR is 1,316 with medical value.

X-rays of 215 patients were analyzed. Algorithm was tested with biorthogonal wavelets, symlets, Daubichies, and reverse biorthogonal wavelets. Our research shows that wavelets of lower orders (less vanishing moments) compress images with more reliability. Bior4.4-wavelet is reliable at low levels of the decomposition, while bior1.1 or bior1.3 are more reliable at higher levels of decomposition (higher CR). The best performance is shown by reverse biorthogonal wavelet—rbio1.3 and Symlet of 2 order (sym2).

6.5 Conclusions

Total number of patients is 215. However, number of images are considerable grater. Except original images, the most important part was to analyze reconstructed images. For every patient, analyzed images were also reconstructions from every level till 14th and for every type of wavelets (different moments, which means different member of the family).

Basic contributions of the work are in determined facts:

- symlets can be used for up to do 14th level of decomposition and obtained CR = 1316;
- biorthogonal wavelets cannot be used always. Their reliability is till 10th level of decomposition and CR = 131.65;
- reverse biorthogonal wavelets can be used till 14th level of decomposition, because there are of medical value when asbestosis is considered,

It is important to point out that data presented is medically valid, because the independency restriction posed by ILO. Algorithm produce 100 % correct images if mentioned wavelets are used for determined levels of decomposition.

Successful compression of pulmonary X-rays is obtained in the research by wavelet transform. Maximal obtained CR was 1316, which means reduction of disk space from 4.5 MB to 3.5 KB.

Tests and experimental work was performed in Matlab. However, system can be automatized by specifying input/output drivers and exporting programing code as C++ project and/or exe-file.

References

1. Pavlidis, T., Steiglitz, K.: The automatic counting of asbestos fibers in air samples. IEEE Trans. Comput. **C-27**(3), 258–261 (1978)
2. Paustenbach, D.J.: Bhopal, asbestos, and Love Canal… how they should affect engineering education. IEEE Technol. Soc. Mag. **6**(1), 9–15 (1987)
3. Petja, B.M., Twumasi, Y.A., Tengbeh, G.T.: The use of remote sensing to detect asbestos mining degradation in Mafefe and Mathabatha, South Africa. In: IEEE International Conference on Geoscience and Remote Sensing, pp. 1591–1593 (2006)
4. Petja, B.M., Twumasi, Y.A., Tengbeh, G.T.: Comparative analysis of reflectance spectroscopy and laboratory based assessment of asbestos pollution in the rehabilitated mining environment, South Africa. In: IEEE International Geoscience and Remote Sensing Symposium, pp. 1246–1249 (2007)

5. Ishizu, K., Takemura, H. et al.: Image processing of particle detection for asbestos qualitative analysis support method-particle counting by using color variance of background. In: SICE Annual Conference, pp. 3202–3207, Tokyo, 20–22 Aug 2008
6. Kawabata, K., Tsubota, Y. et al.: Development of an automatic polarized microscopic imaging system for asbestos qualitative analysis. In: IEEE/ASME International Conference on Advanced Intelligent Mechatronics, AIM 2009, pp. 1671–1676 (2009)
7. Bassani, C., Cavalli, R.M., et al.: Airborne emissivity data to map the urban materials to be checked for maintenance: The asphalt paving and asbestos cement roofing case studies. Joint Urban Remote Sensing Event **2009**, 1–6 (2009)
8. Vujović, M., Vujović, I., Kuzmanić, I.: New technologies and diagnosis of the professional asbestosis. Arch. Environ. Health **49**(3), 251–258 (1998)
9. Vujović, I., Kuzmanić, I.: Histogram analysis of X-ray images and wavelet influence to the contained information. Med. Biol. Eng. Comput. **37**(supp. 2), 1062–1063 (1999)
10. Vujović, I.: Digital image analysis and computer aid in diagnostics of asbestosis (in Croatian). Elektrotehnika **43**(1–2), 17–22 (2000)
11. Vujović, M., Vujović, I., Kuzmanić, I.: The application of new technologies in diagnosing occupational asbestosis. Arch. Environ. Health **54**(4), 245–252 (2003)
12. Vujović, I.: Application of wavelets in biomedical data processing with example in compression of X-rays of occupational asbestosis infected patients. MSc Thesis, University of Split, Faculty of Electrical Engineering, Mechanical Engineering and Naval Arhitecture (2004)
13. Cvitanović, S., Znaor, L.J., et al.: Malignant and non-malignant asbestos-related pleural and lung disease: 10-year follow-up study. Croat. Med. J. **44**(5), 618–625 (2003)
14. Vujović, M.: Standardizing diagnostic criteria for assessment of asbestos- related occupational disease of the lung and pleura. Arch. Environ. Health **46**, 445–449 (1995)
15. Simpson, S.G., Comstok, G.W.: Lung cancer and housing characteristics. Arch. Environ. Health **38**, 248–252 (1983)
16. Akay, M.: Time frequency and wavelets in biomedical signal processing. IEEE Press, New York (1998)
17. Muyshondt, R.A., Mitra, S.: Visual fidelity of reconstructed radiographic images using wavelet transform coding and JPEG. In: 8th IEEE Symposium on Computer-Based Medical Systems, Lubbock, USA (1995)
18. Wang, H., Lai, S.L., Jiang, Y.H.: A comparative study of wavelet used in DICOM image compression. Chin. J. Med. Imaging Technol. **18**(8), 827–829 (2002)
19. Heer, K., Reinfelder, H.E.: A comparison of reversible methods for data compression. In: Proceedings of SPIE "Medical Imaging IV", SPIE, vol. 1233, pp. 354–365 (1990)
20. Said, A., Pearlman, W.A.: An image multiresolution representation for lossless and lossy compression. IEEE Trans. Image Process. **5**(9), 1303–1310 (1996)
21. Calderbank, A.R.; Daubechies, I., Sweldens, W., Yeo, B.L.: Lossless image compression using integer to integer wavelet transforms. In: Proceedings of International Conference on Image Processing ICIP, vol. 1, pp. 596–599. Washington, DC, USA, 26–29 Oct 1997
22. Boles, W.W.: A security system based on human iris identification using wavelet transform. Eng. Appl. Artif. Intell. **11**(1), 77–85 (1998)
23. Grosbois, R.: Image security and processing in the JPEG 2000 compressed domain. PhD Thesis, Université Paris, France (2003)
24. Dai, D.Q., Yuen, P.C.: Wavelet based discriminant analysis for face recognition. App. Math. Comput. **175**(1), 307–318 (2006)
25. Mallat, S.: A Wavelet Tour of Signal Processing: The Sparse Way, 3rd edn. Academic Press, Burlington (2009)
26. Curent Status of DICOM Standard. http://www.dclunie.com/dicom-status/status.html. Accessed 14 Jan 2010
27. Pegasus Imaging Coorporation, Apollo 1.0. http://www.pegasusimaging.com. Accessed 23 July 2006
28. Guidelines for the Use of ILO International Classification of Radiographs of Pneumoconioses. International Labour Office, Geneva (1980)

Part III
Materials Applications

Chapter 7
Strain Measurement in an Aluminium Foam by Means of Digital Image Correlation

Luca Goglio, Marco Peroni and Jakson Manfredini Vassoler

Abstract Metallic foams represent a particular class of materials, characterized by their cellular internal structure, which are receiving a growing interest for lightweight construction and impact absorbers. The inhomogeneity makes it difficult to measure the strain with conventional techniques (e.g. strain gauges); on the contrary, an optical non-contact technique is particularly suitable, taking advantage of the naturally speckled surface. This chapter presents the application of the Digital Image Correlation (DIC) to the study of the response of an aluminium foam subjected to compression. A key aspect of the mechanical characterization of the foam is to evaluate the influence of the density on the sensitivity to the loading rate. The measuring technique encompasses two steps: first the displacements of a set of marker points are tracked using image analysis; then the strains are evaluated by means of a strain-displacement relationship, in which the markers play the role of the nodes in finite element modelling. The results allow for evaluating the evolution of the strains in the material during the compression test.

Keywords Digital image correlation · Metallic foams · Compression testing · Strain measurement

L. Goglio (✉)
Politecnico di Torino, corso Duca degli Abruzzi 24, 10129 Torino, Italy
e-mail: luca.goglio@polito.it

M. Peroni
EC Joint Research Centre, IPSC Institute, ELSA unit, via E. Fermi 2749, 21027 Ispra, VA, Italy
e-mail: marco.peroni@jrc.ec.europa.eu

J. M. Vassoler
Universidade Federal do Rio Grande do Sul, Av. Paulo Gama, 110, Porto Alegre 90040-060, Rio Grande do Sul, Brazil
e-mail: vassoler@mecanica.ufrgs.br

F. R. Leta (ed.), *Visual Computing*, Augmented Vision and Reality 4,
DOI: 10.1007/978-3-642-55131-4_7, © Springer-Verlag Berlin Heidelberg 2014

7.1 Introduction

Full-field, non-contact, optical techniques can add a substantial contribution to materials science for what concerns the measurement of deformation. The main advantage is that the measurement is achieved on a wide region of the sample surface and can be carried out in parallel, while a mechanical testing (e.g. a tension or compression test) is underway. In general terms, these techniques can be classified as interferometric and non-interferometric. In the case of the interferometric technique, the measurement is based on the phase difference (a coherent light source is used) between the light reflected by the surface of the specimen before and after deformation. The technique can detect very small displacements and strains; its main shortcoming is that an elevated level of insulation from vibrations is required. The non-interferometric technique is based on the comparison of the pattern of "speckles" before and after deformation; it is accomplished by processing electronically the related images and, usually, referred to as Digital Image Correlation (DIC). The speckles can be applied (e.g. by spraying a suitable paint) or formed by the natural texture of the specimen surface. This technique cannot be as sensitive as the previous, but is less demanding in terms of experimental setup for what concerns insulation from vibrations and the equipment is less sophisticated. Moreover, the measurement can be performed also with natural light. Thanks to the progress of the electronic equipment, in the last decade both CCD cameras and computers have become more and more powerful and inexpensive so that acquisition and processing of the images is easily achieved. The core of the technique lies mainly in the algorithm adopted to track the speckles in the deformed state and evaluate the displacements, a wide survey of these aspects—and of the features of DIC as well—can be found in [1]. In its two-dimensional formulation DIC requires a single camera, having the plane of its CCD parallel to that of the specimen surface. This setup is suitable as long as the specimen is flat and the out-of-plane motion is negligible; otherwise the more sophisticated 3D DIC, based on a binocular observation, must be applied.

This chapter reports an experience obtained by applying DIC to the measurement of the strain during compression testing of aluminium foam. For this case, contact techniques would be unsuitable, due to the porous nature of the specimen surface and the inhomogeneity of the material. On the contrary, the optical technique is fit to the purpose, taking also advantage from the naturally speckled surface of the specimens. Moreover, since displacements and strains are large, the lower resolution of DIC with respect to interferometry does not constitute a drawback. The text is organised in the following way: first, a brief description of the aluminium foams is presented, especially for what concerns the assessment of their mechanical properties; second, the results of the compression tests carried out in the work are reported; then, the application of DIC is described.

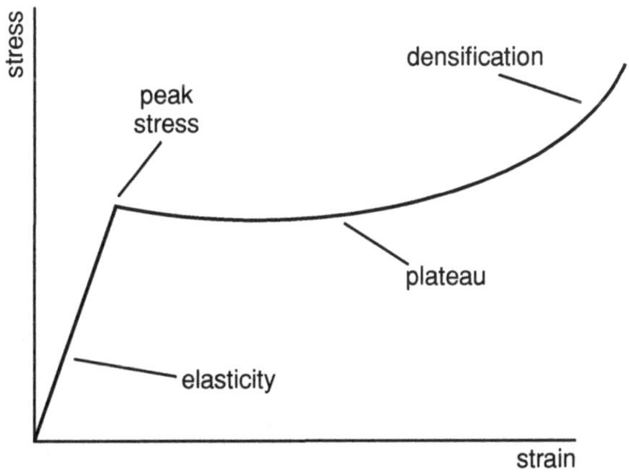

Fig. 7.1 Schematic representation of the behaviour of a foam under compression

7.2 Aluminium Foams

Although metal foams [2] are getting popular, some related essential information is given here for the benefit of a reader inexperienced with this class of materials. In short terms, in a metal foam the metallic matrix is interrupted by cavities, purposely generated by the manufacturing processes. This structure is obtained in several ways, the most used are: blowing gas in the molten metal, introducing a chemical agent which produces bubbles in the molten metal, consolidating metal powder containing a foaming agent, casting the metal in a mould containing a wax or polymer precursor of the cavities (protected by a coating of mould material). According to the technology used, the cavities can be communicating or separate, the resulting structures are called "open-cell" or "closed-cell" respectively. Although several metals have been used to produce foams, the most important applications are given by aluminium foams, in which the intrinsic lightness of the metal is emphasised by the cellular structure; the result is an apparent density of 2–300 kg/m^3. Aluminium foams are used to produce lightweight components in aerospace applications, cores of sandwich panels, etc. The cellular structure exhibits high thermal conductivity and is effective in managing vibrations or noise, since it offers better stiffness (which increases the natural frequencies) and damping than a solid panel of same mass per unit surface. A peculiar application of the foam is to dissipate a considerable amount of energy when is crushed; therefore it is used to produce impact absorbers, in form of pads or by filling the cavities of thin-walled tubular elements. The stress-strain response of a foam under compression is depicted schematically in Fig. 7.1. Under increasing load, the graph exhibits first an elastic rise, until a peak value is attained; then the stress stabilizes at an approximately constant value (plateau); finally the stress increases noticeably, due to the closure of the voids deforming in compression (densification). The actual behaviour is somewhat more complicated, as some inelastic phenomena occur already in the initial rise.

Due to the practical interest, and to the relative novelty of this class of materials, much research work has been spent in the last two decades to investigate their mechanical properties and response under several types of loading. A short survey (limited to years from 2,000 onwards) is reported in the following.

Numerous researches have contributed to establish a wide experimental knowledge of the mechanical properties. In [3] the deformation mechanisms of closed-cell foams is analysed, accounting for cases of distortion alone or accompanied by rotation and shear; it is also remarked that the deformation initially appears in spots, then spreads out over bands. The evolution from localised deformation to compaction bands is studied in [4]. In [5] the failure mechanism of closed-cell foams is studied and it is noticed that the cracks can nucleate and propagate through the brittle precipitates embedded in the walls. Regarding the effect of the cell size, in [6] it is found that, comparing foams of equal density, the strength is higher in case of larger diameter of the cell, due to the higher wall thickness.

More complex testing conditions have been adopted to assess the response of the foams under multiaxial loading; the obtained results support the definition of empirical models to predict the yield and post-yield behaviour, and validate the models proposed on a theoretical basis. Results of this kind are found in [7] and [8] by means of a hydraulic apparatus, while in [9] a biaxial testing machine is used. Maybe the largest experimental database on the mechanical behaviour of the foams, established to support the finite element code LS-DYNA, is described in [10]. Other contributions have assessed the capability of known formulae to predict the moduli of the foams [11], or have proposed regression models [12]. A model to predict elastic modulus and plateau stress is proposed and validated in [13].

As stated previously, a prominent use of the aluminium foams is to increase the absorption of energy in impact conditions. Therefore, many works have been dedicated to the effect of the strain rate on the adhesive performance. Several of the studies are based on experiments performed by means of the Split Hopkinson Pressure Bar (SHPB), which is required to obtain strain rates above 10 s^{-1}. In [14] the dependence is assessed over a wide range of strain rate, of relatively low values, and it is noticed that at low strain rate the deformation is localised in some zones, while at higher strain rate deformation bands appear and the effect of the inertia becomes predominant. In [15] it is found that the dependence on the strain rate is higher for the foams with higher density, and the reason is ascribed to the gas flowing through the orifices in the wall during the progressive failure. Conversely, in [16], for a closed-cell foam, the plateau stress is found to be almost insensitive to the strain rate but is affected by the density. In [17] three different foams are compared; the dependence on the strain rate is significant for one foam and partial for another, conversely the third foam is insensitive. Foams obtained by sintering metal powder have been studied in [18], also in this case it has been found that the mechanical properties depend on strain rate and density. The response of a foam to static and dynamic loading, considering different densities, is investigated in [19] and the evidence is that the dependence on the strain rate is higher in case of high density. This is also the remark that can be made as an interpretation of the general behaviour, although some dissimilar results have been found (e.g. [20], in which only evidence for the effect of the density is found).

It is obvious that, to assess the mechanical properties of a foam, in several cases an effective measurement of the strain is needed. Only a mean value of the axial strain can be obtained from the displacement of the crosshead (or equivalent piece of equipment), whilst no information on the transverse strain is available. Moreover, for an inhomogeneous material as a foam, the local deformation values can be very far from the mean. For the reasons already cited at the end of Sect. 7.1, DIC appears as the ideal technique to measure the strain field; thus, for instance, it has been used in [3–5]. Using a suitable high speed camera, measurement under high rate of deformation can also be performed, as done in [21].

7.3 Mechanical Tests

The discussions reported in this chapter are based on experiments [22] carried out on Alporas®, an aluminium foam produced with the technique of generating bubbles in the molten metal by means of TiH_2 [23]. The material has been cut in cubical samples having side of 50 mm, which is one order of magnitude larger than the typical size of the cavities, as usually recommended. The nominal density is 250 kg/m^3; however, due to the inhomogeneity related to the production process, the value is lower in the middle and higher at the edges of the plates in which the material is manufactured. Since, in practice, each sample had a different density, in the range 232–306 kg/m^3, to study the influence of the density the specimens were grouped in three classes of density, termed "low" (235.4–238.5 kg/m^3), "medium" (248.8–252.8 kg/m^3) and "high" (270.1–272.5 kg/m^3). The classes defined in this way contained a sufficient number of samples (about ten), enough similar each other and different from those of the other classes. The reason for defining these classes was to have specimens of practically equal density to be tested under different velocities.

All compression tests have been carried out with a servo-hydraulic machine of 100 kN maximum load and 100 mm/s maximum crosshead speed (Dartec HA100), equipped with platens instead of the usual jaws, as shown in Fig. 7.2. The testing velocities have been chosen one order of magnitude different each other, 0.25 (low), 2.5 (intermediate), 25 (high) mm/s; the corresponding strain rates referred to the initial specimen height are respectively $5{\cdot}10^{-3}$, $5{\cdot}10^{-2}$, $5{\cdot}10^{-1}$ s^{-1}. Each test has been replicated three times under the same nominal conditions (density and velocity). The total length of the compressing stroke is 40 mm, corresponding to 80 % strain.

The load-displacement curves recorded during the tests are shown in Fig. 7.3, each diagram corresponds to one velocity as labelled on top. Although, for each velocity, the curves are not much different, from the plateau onwards they appear stacked according to the density, and this appears more evident increasing the velocity. These observations are consistent, although in this case the density varies in a narrower range, with the findings reported in [15].

Fig. 7.2 Compression testing of a foam block

7.4 Optical Strain Measurement

7.4.1 Displacements

The strain distribution in the specimen during the test was obtained by means of DIC, as discussed in the introduction. Regarding the application of DIC to the measurement on foams undergoing deformation, it must be emphasized that the surface of the foam presents naturally different tones of gray, due to the pores, that can act as a pattern of speckles for the measurement. If, on the one hand, due to the typical cell size (some mm) the resolution of the measurement cannot be very fine, on the other hand the displacements involved in the phenomenon are large, and this facilitates the measurement. On the surface of the specimen a grid of points, termed *markers*, is first identified. Then, a rectangular zone around each marker, the *subset*, is considered with its tones of gray. The position of the subsets is tracked during the mechanical test and this gives the displacements of the markers, as shown schematically in Fig. 7.4. In this scheme the subsets are chosen aligned on their outlines, it can be remarked that they could also be chosen overlapping or separated.

To evaluate the displacements, the procedure examines the gray scale over the image, searching the similarity of the subsets. The position of the marker is that which maximizes the normalized cross-correlation (NCC)

$$r(i,j) = \frac{\sum_{m=1}^{M_a} \sum_{n=1}^{N_a} A(m,n) \cdot B(m+i, n+j)}{\sqrt{\sum_{m=1}^{M_a} \sum_{n=1}^{N_a} A(m,n)^2 \cdot \sum_{m=1}^{M_a} \sum_{n=1}^{N_a} B(m+i, n+j)^2}} \tag{7.1}$$

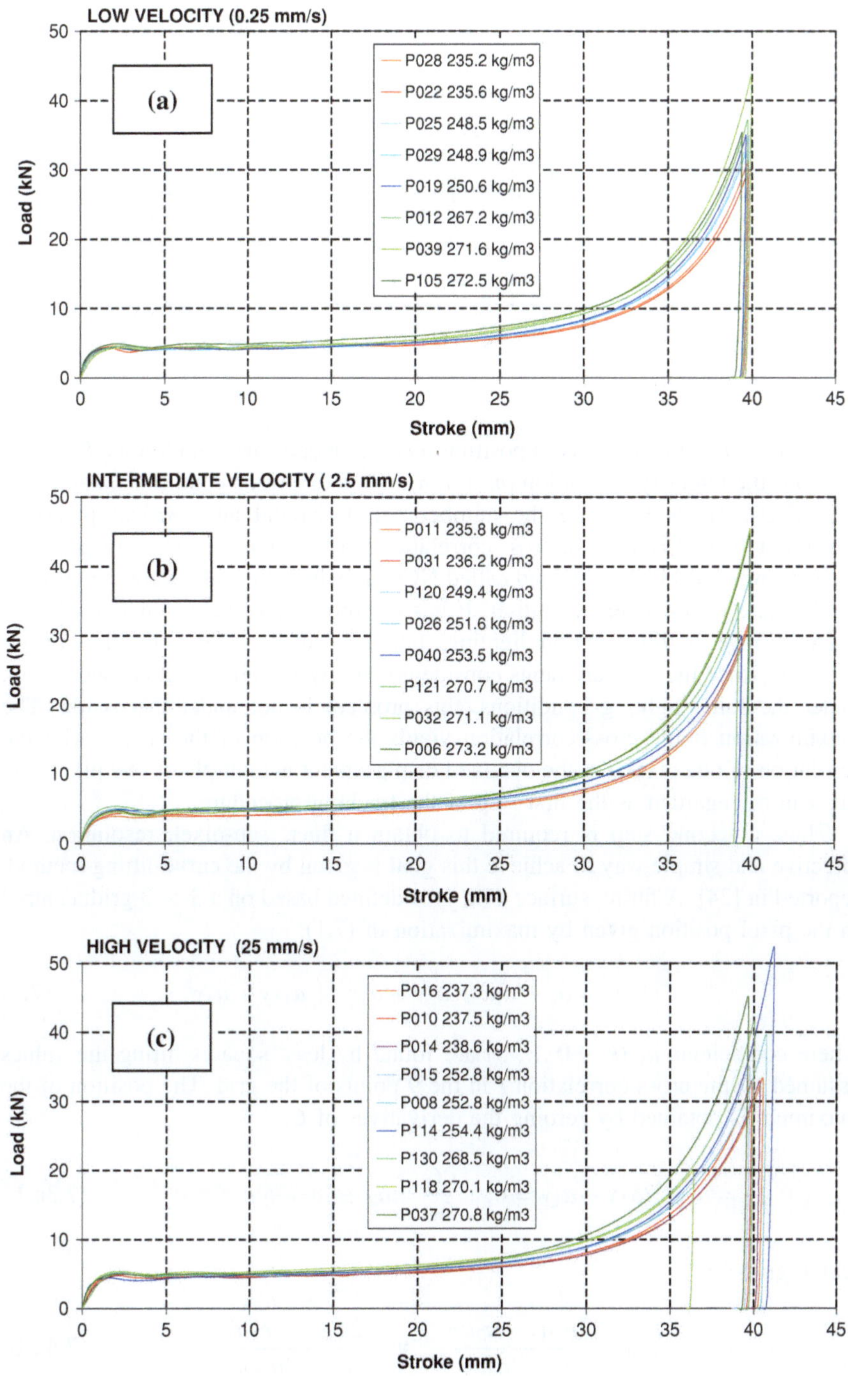

Fig. 7.3 Load-displacement curves recorded at: **a** low velocity; **b** intermediate velocity; **c** high velocity

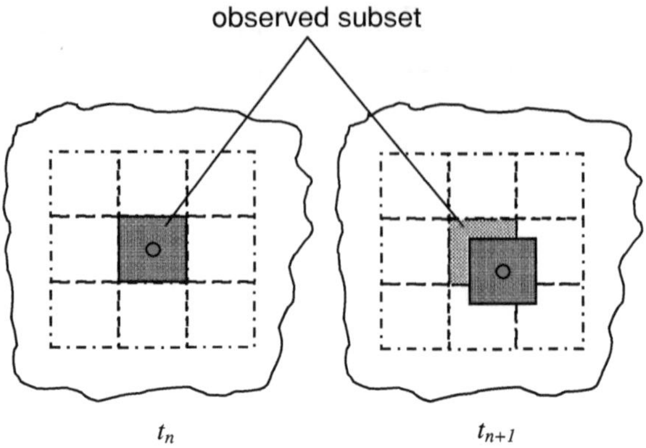

Fig. 7.4 DIC: evolution of a subset during deformation from instant t_n to instant t_{n+1}

where $A(m, n)$ is the intensity at position (m,n) in undeformed conditions, $B(m + i, n + j)$ is the intensity at position $(m + i, n + j)$ in deformed conditions, i and j are the offsets, M_a and N_a are the numbers of horizontal and vertical pixels in the subset. This definition of cross correlation is equivalent to that reported in the second row of Table 1 in [1] and called NCC, in which the values are weighed by their square means over the subset. It has the property of being unaffected by a linear change in illumination lighting, but it is affected by a lighting offset; however, since the measurements considered here were carried out consecutively, under the same lighting conditions, this problem is not under discussion. The maximization if the cross-correlation yields the position of the subset with the resolution of one pixel, i.e. the obtained displacement is multiple of one pixel, and this can be regarded as the first step of the tracking procedure.

Thus, a second step is required to obtain a finer, sub-pixel, resolution. An effective and simple way to achieve this goal is given by the curve-fitting method, reported in [24]. A fitting surface $C(x, y)$ is defined based on a 3 × 3 grid, centred in the pixel position given by maximization of (7.1):

$$C(x, y) = a_0 + a_1 x + a_2 y + a_3 x^2 + a_4 xy + a_5 y^2 \tag{7.2}$$

where coefficients a_k ($k = 0,\ldots, 5$) are found by least squares fitting the values assumed by the cross-correlation r in the 9 points of the grid. The position of the maximum is obtained by zeroing the derivatives of C:

$$\frac{\partial C}{\partial x} = a_1 + 2a_3 x + a_4 y = 0; \quad \frac{\partial C}{\partial y} = a_2 + 2a_5 y + a_4 x = 0 \qquad (7.3\text{a, b})$$

which gives:

$$x = \frac{2a_1 a_5 - a_2 a_4}{a_4^2 - 4a_3 a_5}; \quad y = \frac{2a_2 a_3 - a_1 a_4}{a_4^2 - 4a_3 a_5} \qquad (7.4\text{a, b})$$

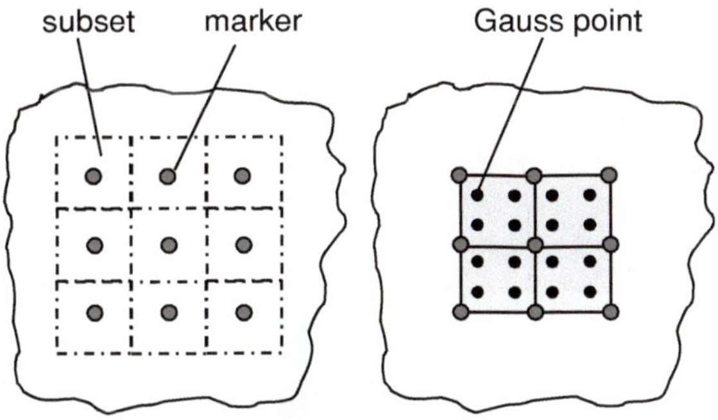

Fig. 7.5 Strain evaluation: markers interpreted as nodes of a finite element mesh and related Gauss points

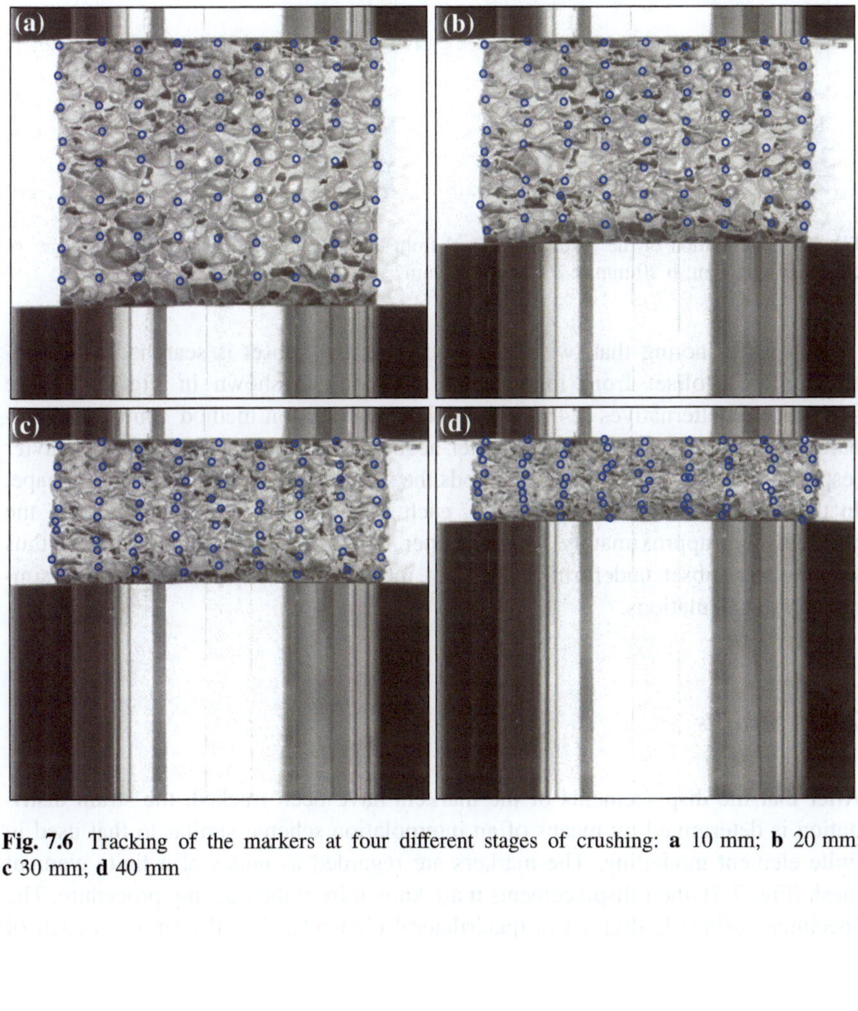

Fig. 7.6 Tracking of the markers at four different stages of crushing: **a** 10 mm; **b** 20 mm; **c** 30 mm; **d** 40 mm

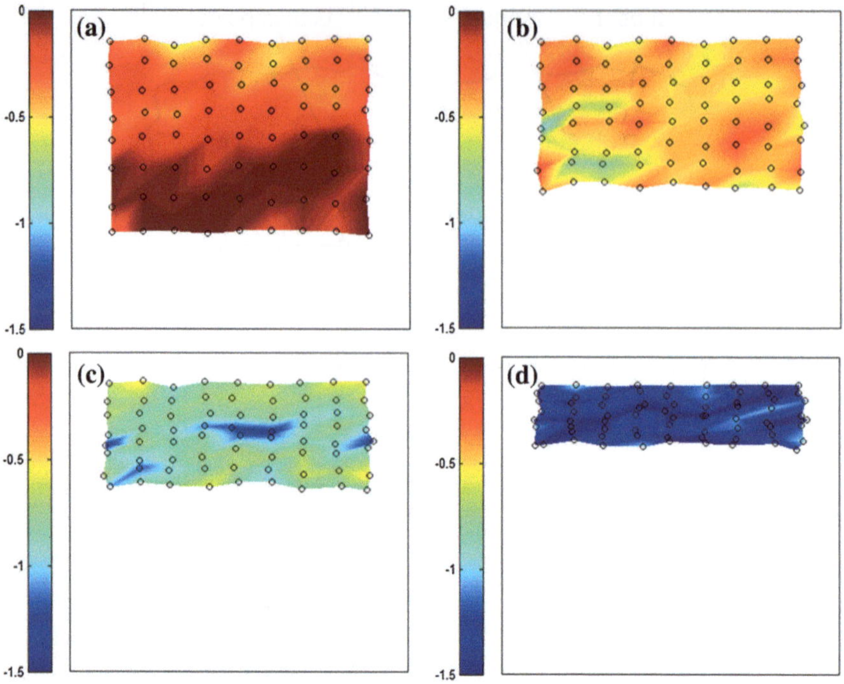

Fig. 7.7 Distribution of the *y* strain obtained from the optical measurement at four stages of crushing: **a** 10 mm; **b** 20 mm; **c** 30 mm; **d** 40 mm

It is worth noting that, with this procedure, the subset is searched as a rectangular area, offset from the original position, as shown in Fig. 7.4. More sophisticated alternatives [24] are the "Newton-Raphson method" and "gradient-based method", which achieve a better accuracy but are computationally heavier (especially the former); in these methods the subset is tracked in a deformed shape. In the measurements described here, each image is not too different from the previous one (approximately 10 images per 1 mm of stroke are considered), thus keeping the subset undeformed does not introduce an important error and simplifies the calculations.

7.4.2 Strains

After that the displacements of the markers have been tracked, the strain distribution is determined by means of an interpolation scheme similar to that used in finite element modelling. The markers are regarded as nodes of a finite element mesh (Fig. 7.5), their displacements **u** are known from the tracking procedure. The specimen surface is divided in quadrilateral elements, thus the strain in each of

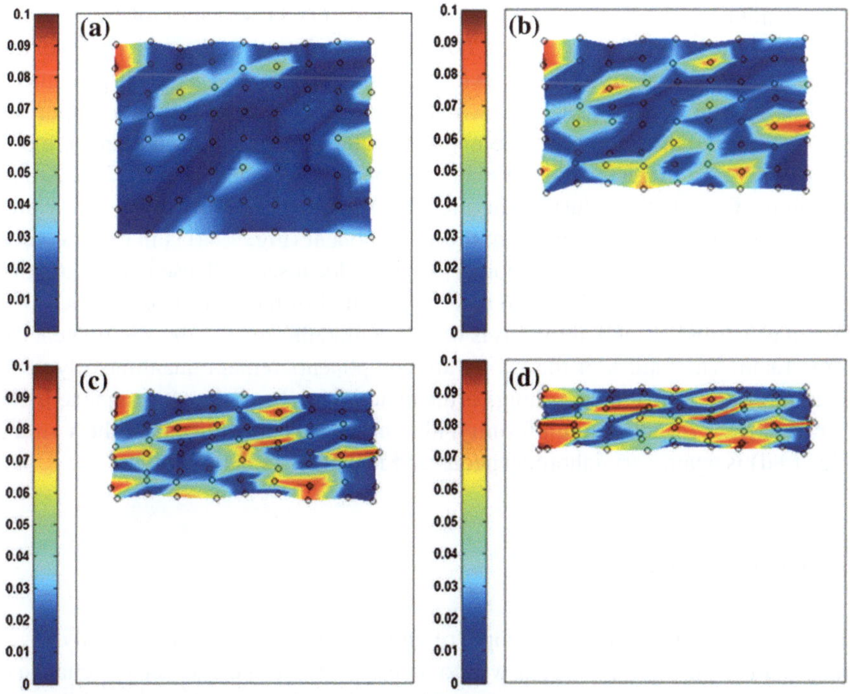

Fig. 7.8 Distribution of the x strain obtained from the optical measurement at four stages of crushing: **a** 10 mm; **b** 20 mm; **c** 30 mm; **d** 40 mm

them can be calculated via the well known strain-displacement matrix **B**. Since **B** is usually defined with the natural coordinates (ξ, η), a mapping is required to the physical coordinate (x, y); this is accomplished with the usual shape functions of finite elements. In conclusion, the strains **ε** are calculated as:

$$\boldsymbol{\varepsilon} = \mathbf{B}\mathbf{u} \tag{7.5}$$

A peculiarity of the considered case is that the foam undergoes very large strains. This has the detrimental consequence on the measurement technique that the features of the speckle can change to a great extent; thus, correlating the deformed to the undeformed shape can be difficult or even impossible. Again, as reported at the end of Sect 7.4.1 regarding the search of the subset as undeformed rectangle, the remedy is considering several images during the evolution of the phenomenon and correlating each image to the previous one instead of the initial one.

The procedures for tracking the markers and calculating the strains, performing all the related calculations, have been implemented in MATLAB®. Figure 7.6 shows, at four different stages of crushing (10, 20, 30, 40 mm), the markers superposed to the images of the deformed specimen. Figures 7.7 and 7.8 show, respectively, the distribution of the longitudinal (y) and transverse (x) strains, corresponding to the four crushing stages of Fig. 7.6. Observation of the contour plots gives an interesting insight on how the deformation in the material evolves during the test. The process is, obviously, governed by the axial displacement imposed by the lower platen, moving upwards. The longitudinal component (Fig. 7.7) is, in absolute value (the negative values indicates compression), more than one order of magnitude larger than the transverse component (Fig. 7.8). At the beginning of the crushing the y strain (Fig. 7.7a) distribution grows in absolute value from top to bottom of the specimen; the x component (Fig. 7.8a) is in general very small except for the bottom left corner, where a localised collapse occurs. As the crushing reaches (Figs. 7.7b, 7.8b) and exceeds half of the total stroke (Figs. 7.7c, 7.8c), the distribution of both strains is very irregular, this reflects the inhomogeneity of the foam. The values of the two strain components remain quantitatively very different. At the end of the crushing, as the material has been almost compacted, the y strain (Fig. 7.7d) tends to uniformity, although bands are noticeable; the x strain (Fig. 7.8d) is again very inhomogeneous and its values remain small.

7.5 Conclusions

The chapter has presented an example of application of the DIC to the study of an aluminium foam. The main goal was to assess if strain rate and density changes, considering for the former the range of low values 10^{-3}–10^{-1} s^{-1}, and for the latter the intrinsic change of values from centre to periphery (232–306 kg/m^3), affect the properties of the foam. Compressive tests have been carried out on cubic specimens of material; it is observed that the effect on the response (expressed as force-displacement diagram) is noticeable, although small.

Digital Image Correlation has been applied to study the strain field in the sample during the compression tests. The displacements of the marker points are detected, then the strains are calculated by adopting an interpolation of the same type as that used in finite element modelling. The results given by the application of the DIC to this kind of material are valuable to assess the strain distribution in the sample surface, which would be impossible to obtain with other experimental contact techniques (e.g. strain gauges).

References

1. Pan, B., Qian, K., Xie, H.M., Asundi, A.: Two-dimensional digital image correlation for in-plane displacement and strain measurement: a review. Meas. Sci. Technol. **20**, 1–17 (2009)
2. Ashby, M.F., Evans, A.G., Fleck, N.A., Gibson, L.J., Hutchinson, J.W., Wadley, H.N.G.: Metal Foams: A Design Guide. Butterworth-Heinemann, Oxford (2000)
3. Bastawros, A.F., Bart-Smith, H., Evans, A.G.: Experimental analysis of deformation mechanisms in a closed-cell aluminum alloy foam. J. Mech. Phys. Solids **48**, 301–322 (2000)

4. Issen, K.A., Casey, T.P., Dixon, D.M., Richards, M.C., Ingraham, J.P.: Characterization and modeling of localized compaction in aluminum foam. Scripta Mater. **52**, 911–915 (2005)
5. Amsterdam, E., De Hosson, J.Th.M., Onck, P.R.: Failure mechanisms of closed-cell aluminum foam under monotonic and cyclic loading. Acta Mater **54**, 4465–4472 (2006)
6. Yu, H., Guo, Z., Li, B., Yao, G., Luo, H., Liu, Y.: Research into the effect of cell diameter of aluminum foam on its compressive and energy absorption properties. Mater. Sci. Eng. A **454–455**, 542–546 (2007)
7. Deshpande, V.S., Fleck, N.A.: Isotropic constitutive models for metallic foams. J. Mech. Phys. Solids **48**, 1253–1283 (2000)
8. Ruan, D., Lu, G., Ong, L.S., Wang, B.: Triaxial compression of aluminium foams. Compos. Sci. Technol. **67**, 1218–1234 (2007)
9. Öchsner, A., Kuhn, G., Grácio, J.: Investigation of cellular solids under biaxial stress states. Exp. Mech. **45**, 325–330 (2005)
10. Hanssen, A.G., Hopperstad, O.S., Langseth, M., Ilstad, H.: Validation of constitutive models applicable to aluminium foams. Int. J. Mech. Sci. **44**, 359–406 (2002)
11. Mondal, D.P., Ramakrishnan, N., Suresh, K.S., Das, S.: On the moduli of closed-cell aluminum foam. Scripta Mater. **57**, 929–932 (2007)
12. Edwin Raj, R., Daniel, B.S.S.: Structural and compressive property correlation of closed-cell aluminum foam. J Alloy Compd. **467**, 550–556 (2009)
13. Konstantinidis, ICh., Papadopoulos, D.P., Lefakis, H., Tsipas, D.N.: Model for determining mechanical properties of aluminum closed-cell foams. Theor. Appl. Fract. Mech. **43**, 157–167 (2005)
14. Paul, A., Ramamurty, U.: Strain rate sensitivity of a closed-cell aluminum foam. Mater. Sci. Eng. A **281**, 1–7 (2000)
15. Dannemann, K.A., Lankford Jr, J.: High strain rate compression of closed-cell aluminium foams. Mater. Sci. Eng. A **293**, 157–164 (2000)
16. Ruan, D., Lu, G., Chen, F.L., Siores, E.: Compressive behaviour of aluminium foams at low and medium strain rates. Compos. Struct. **57**, 331–336 (2002)
17. Montanini, R.: Measurement of strain rate sensitivity of aluminium foams for energy dissipation. Int. J. Mech. Sci. **47**, 26–42 (2005)
18. Yi, F., Zhu, Z., Zu, F., Hu, S., Yi, P.: Strain rate effects on the compressive property and the energy-absorbing capacity of aluminum alloy foams. Mater. Charact. **47**, 417–422 (2001)
19. Edwin Raj, R., Parameswaran, V., Daniel, B.S.S.: Comparison of quasi-static and dynamic compression behavior of closed-cell aluminum foam. Mater. Sci. Eng. A **526**, 11–15 (2009)
20. Hall, I.W., Guden, M., Yu, C.-J.: Crushing of aluminum closed cell foams: density and strain rate effects. Scripta Mater. **43**, 515–521 (2000)
21. Reu, P.L., Miller, T.J.: The application of high-speed digital image correlation. J. Strain Anal. Eng. **43**, 673–688 (2008)
22. Goglio, L., Vassoler, J.M., Peroni, M.: Measurement of longitudinal and transverse strain in an aluminium foam. Materialwiss. Werkstofftech. Mater. Sci. Eng. Technol. **42**, 342–349 (2011)
23. Miyoshi, T., Itoh, M., Akiyama, S., Kitahara, A.: ALPORAS aluminum foam: production process, properties and applications. Adv. Eng. Mater. **2**, 179–183 (2000)
24. Pan, B., Xie, H.M., Xu, B.Q., Dai, F.L.: Performance of sub-pixel registration algorithms in digital image correlation. Meas. Sci. Technol. **17**, 1615–1621 (2006)

Chapter 8
Analysis of the Surface of Parts Injected with ABS and Treated with Acetic Acid: Influence of Process Settings

R. P. Bom and E. H. Murakami

Abstract This study examined the influence of process settings of Acrylonitrile Butadiene Styrene (ABS) injection molding on the surface residual stress generated during the filling, packing, and cooling phases. The thickness of layers of Cu–Ni–Cr electro deposited by optical microscopy was determined. Several specimens were injection molded with ABS and treated with different process parameters. Specimen surfaces were treated with glacial acetic acid for 2 min at room temperature. Analysis of surfaces by electron microscopy was performed. This enabled the identification of the influence of injection parameters on surface morphology after treatment with acid. The correlation of results with the surface residual stress obtained through simulation was determined. The method used for determining surface residual stress which damages the electro deposit process originated from injection mold was experimentally determined upon the absence or presence of surface bleaching. Upon the interaction of statistical, numerical, and experimental tools, an equation and a response surface model were established. From this surface, following an isoline of residual stress, it was experimentally demonstrated that it is possible to obtain specimens with non-bleached surfaces with low injection time and packing pressure equal to injection pressure.

Keywords Residual stress · Acetic acid · Injection settings · ABS

R. P. Bom (✉)
Mechanical Engineering Department, Center for Science and Technology—CCT, University of the State of Santa Catarina—UDESC, Campus Universitário Avelino Marcante, B. Bom Retiro, CEP, Joinville, SC 89223-100, Brazil
e-mail: ricardo.bom@udesc.br

E. H. Murakami
Whirlpool S.A. Compressor Plant, Joinville, SC, Brazil

F. R. Leta (ed.), *Visual Computing*, Augmented Vision and Reality 4,
DOI: 10.1007/978-3-642-55131-4_8,

8.1 Introduction

Injection molding is an important process for the bathroom fixture industry. The deposition of metal layers on plastic by electroplating for the purpose of obtaining chromed parts is one choice for a final decorative finish [1, 2]. Currently, during the optimization phase of injection process settings, industries can control residual stress induced in the polymer during the injection process. These stresses can cause deformation and even breakage of parts when these are under the influence of particular environments where plastic parts are used. At low magnitudes of residual stresses, small micro fractures and deformations may occur in layers of materials which are very close to the surface (<50 μm) [3]. When there are major residual stresses, significant warping, cracks, and fractures may occur in the part [4]. In the case of Acrylonitrile Butadiene Styrene (ABS) injection molded parts which subsequently undergo a galvanization process, the condition of residual stresses on the surface is one of the most important factors, directly influencing the quality of the polymer-metal bonding [1, 2]. One of the quality problems is insufficient bonding between the plastic part and the metal coat, as a result of residual stresses present in the plastic part [5]. Several studies have been carried out to measure residual stresses of injection-processed polymers, amongst which "layer removal" [6] and "hole-drilling" [7] are unable to measure stress too close to surface [3]. One technique which is able to identify surface stress levels is "chemical probe" [8], where the relationship between stress and deformations or cracks for different combinations of polymer-aggressive environment is studied in order to compare results with reference values. Upon the application of numerical methods to simulate the 3D flow of a viscoelastic material, residual stresses induced in a virtual model were obtained. Chang et al. compared these stresses calculated for the surface with a part treated with acetic acid and injected in the same manner. This comparison showed promising results [9]. Since thermoplastics used are not conductive, chemical deposition of a conductive layer serving as a base for other electrodeposited layers is required. In the case of ABS, butadiene is oxidized, resulting in several micro cavities and polar groups bonded to the rigid part of ABS. The noble metal Palladium is bonded onto the polar groups, followed by a small deposition and electrolytic nickel. At this stage the porosity is filled, thus anchoring the layer deposited. The subsequent stages are: deposition of copper, nickel, and a small layer of chromium [10]. Another study assesses the concentration effect of conditioning baths on the morphology of surfaces to be metallized and the use of different acids in the process of surface treatment [11].

8.2 Materials and Methods

In this study, several specimens were injected with different process settings in order to carry out treatment with glacial acetic acid and perform surface analysis through scanning electron microscopy (SEM). Analysis and measurement of the influence of injection parameters on treated surface morphology and correlation with surface residual stress determined through simulation were thus obtained. For the injection of specimens, the Tairilac-AG-12AO copolymer ABS supplied by Formosa Chemicals & Fibre Corporation was used. The ABS was dried in an oven at 80 °C for 8 h before processing. Injection equipment used was a Battenfeld model 250 plus with a 22 mm screw, shot weight 34.5 g PS and clamping force 25t. Glacial acetic acid 99 % purity was used to identify the presence of residual stress on the surface of the injected parts. The following experiment was carried out: immersion in acid for 2 min [12] at room temperature followed by washing in running water, drying with compressed air and subsequent SEM analysis at magnifications of 50, 100, 200, and 2,000. Five specimens from each test were obtained. After repetitive trials, only one specimen from each experiment was analyzed through SEM. An optical microscope of 200 magnification with a digital camera was used to determine the layer deposited.

8.3 Results and Discussion

8.3.1 Analysis of One Part from the Assembly Line

In order to carry out flow rate simulations with the software Moldflow version MPI 5.0 [13], a finite element model was developed, as shown in Fig. 8.1. Details of the mold, such as part dimension, mesh grade, number of elements and nodes, and process input settings for the Moldflow program can be found in a master science thesis by Murakami [2]. The part analyzed is an ASTM D 638, type I tensile shear test specimen. Material used for simulation is an injected ABS counter type grade GG-22 from BASF.

Twenty-seven simulations were carried out to determine variations of Surface Residual Stresses.

A centrally located mesh node of the specimen was chosen to determine surface residual stresses, as shown in the surface residual stresses curve of Fig. 8.1. The surface residual stresses curve starts on one side of the specimen and reaches the opposite surface. Figure 8.1 shows the point of the curve where residual stress tension was determined.

Figure 8.2 shows the layers deposited onto the surface of an ABS injected part. The effects of butadiene oxidation on the surface of the part, the incorporation of the metal Palladium, or the electrolytic nickel in this interface cannot be observed.

Optical microscopy enables the measurement of the copper layer thickness as 17 μm and nickel thickness as 12 μm. Chrome layer deposited onto nickel is lower than 1 μm. Total thickness deposited is around 30 μm. These figures are in the range suggested by the literature [10]. Figure 8.3 shows a photograph of the surface of a part from the assembly line to be chromed. This part was treated with acetic acid as described in the trial procedure. In the photograph, one can see that some areas are bleached whereas others are not.

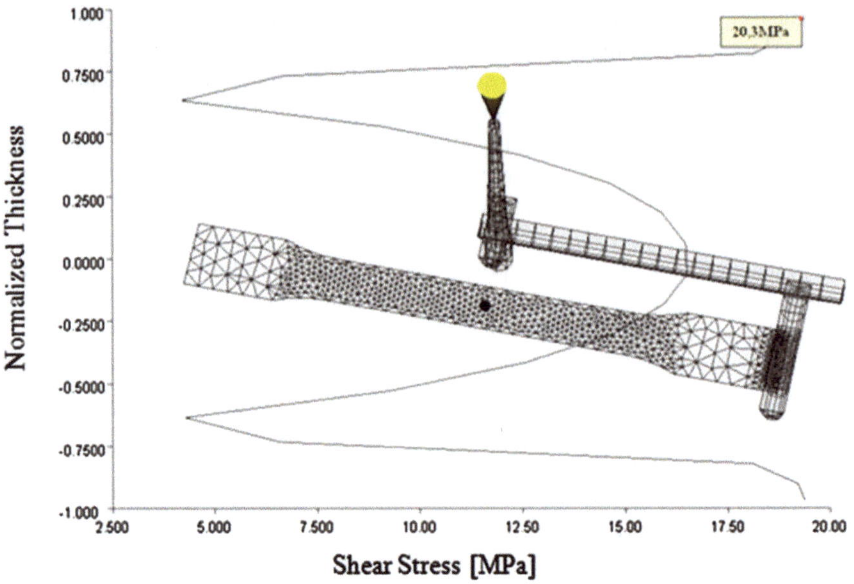

Fig. 8.1 Finite element model type midplane of the specimen with the distribution channels and residual tension curve [2, 13]

Fig. 8.2 Determination of thickness of electro deposited metal layers in an ABS injected part

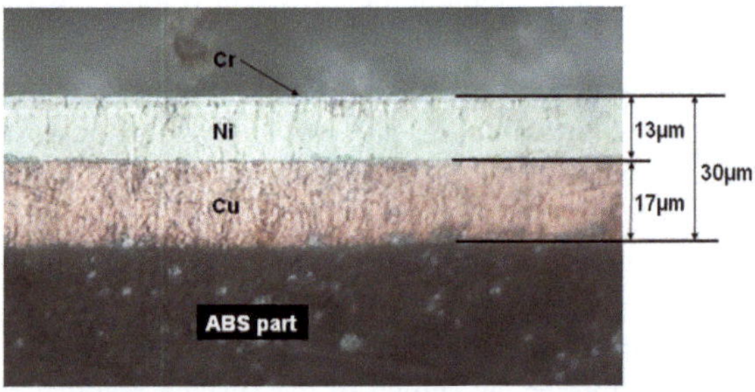

Bleaching caused by the attack of the acid is an indication of the presence of residual stress on the part surface. Electron microscopy analysis has shown (Fig. 8.4a and b) that wherever acid attack was present the surface presented fissures in the shape of micro cavities.

The area attacked was observed through SEM 100 magnification. Micro cavities showed a maximum estimated length of 110 µm and a maximum estimated width of 28 µm. Those areas with reduced residual stress were not susceptible to the attack, thus nothing was observed through a 50 magnification (Fig. 8.4b). Technically speaking, this surface is not adequate for the metallization process. The part shown in Fig. 8.3 is always used in the assembly line and due to the presence of several grooves on the side opposite the surface indicated, the morphology of the attack surface from a simpler part and in the shape of a specimen was chosen.

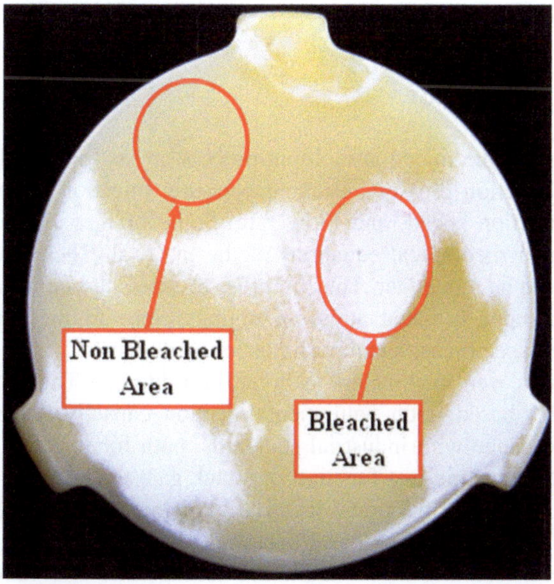

Fig. 8.3 ABS part treated with glacial acetic acid, areas with different residual stresses indicated [provided by DOCOL Faucets]

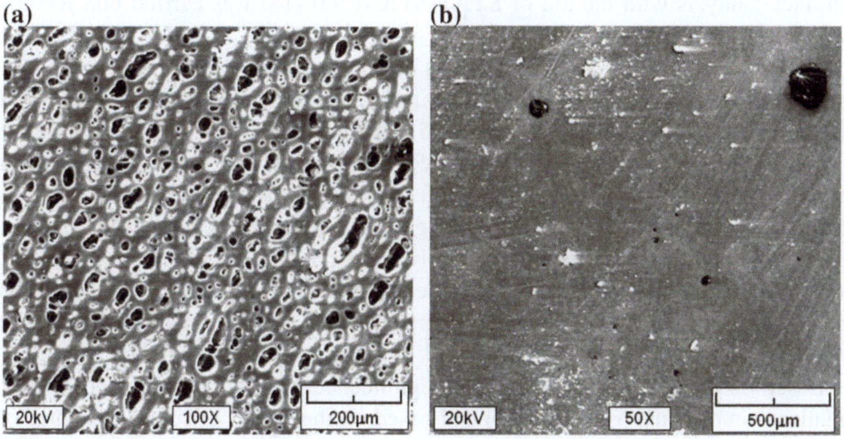

Fig. 8.4 **a** Bleached area magnified 100 times (SEM), presence of micro cavities. **b** Area not affected by acid, magnified 50 times

8.3.2 Statistical Analysis

8.3.2.1 DOE1 Experiment

Main parameters of the injection molding process are as follows: Injection time, Packing time, Injection pressure, Packing pressure, Injection temperature, Mold temperature, Injection speed, and Flow rate. Some of these parameters can be correlated through rheological equations of the material. Therefore, their use in this chapter would be redundant. Thus, in order to assess the influence of injection process settings upon residual stress in regions close to the surface of ABS injection molded specimens, a composite experiment named DOE1 was designed. Parameters chosen were parameters as given in Table 8.1. The limits of process settings were set based upon equipment injection capacity. Also, temperature range is in accordance with industrial standards, both for injection and for mold temperature. Packing time is a function of attack gate thickness.

Experiment DOE1 is a fractional factorial experimental design 2(5–1) totaling 27 settings (S). This experiment has 16 settings plus two axial points for each of the 5 parameters, resulting in 10 additional settings and one central point also considered a setting. These settings can be seen in Table 8.2. Response variables are residual tension values shown in the last column, which were determined by simulation and obtained as indicated in Fig. 8.1.

In order to identify the most relevant parameters for the DOE1 experiment, a variance analysis with the aid of STATISTICA 9.0 [14] was carried out. Results are given in Table 8.3.

The first column shows the parameters being assessed, where numbers stand for the following parameters: (1) Injection Time, (2) Injection Temperature, (3) Mold temperature, (4) Packing Pressure, and (5) Packing Time. The letters indicate the linear (L) and square (Q) (nonlinear) effect of the parameter being assessed upon the variance of the response variable Surface Residual Stresses. The other columns show values of sums of squares (SS), degree of freedom (df), mean square (MS), F0 distribution, and p-value, respectively. The parameters Injection Time (Q), Mold Temperature (Q), and Packing Time (Q) do not show a relevant influence on the response variable and thus may be disregarded. A new variance analysis can be found in Table 8.4.

Based on the results above and on a significance level of 0.05 (chance of error smaller than or equal to 5 % [1, 2]) and based on the values of sums of squares of SS parameters can be classified by degree of influence on Surface Residual Stresses. Injection Time, and Packing Pressure were shown to account for 76.5 % of the variance of Surface Residual Stresses. Other settings did not show influence greater than 3 % each, as can be seen in Fig. 8.5.

Upon the use of the statistical software, a response surface correlating Surface Residual Stresses with the two main parameters identified in the DOE1 experiment (Injection Time and Packing Pressure) was obtained.

Table 8.16 DOE1 parameters and experiment levels

Parameter	Level		
	Minimum	Middle	Maximum
Injection time (s)	0.5	10	20
Injection temperature (°C)	220	245	270
Mold temperature (°C)	40	50	60
Packing pressure (MPa)	30	60	90
Packing time (s)	7	11	15

Table 8.2 DOE1 central composite design settings

S	$2^{(5-1)}$ central composite design—27 settings					
	Injection time (s)	Injection temperature (°C)	Mold temperature(°C)	Packing pressure (MPa)	Packing time (s)	Residual stress (MPa)
1	0.5	220	40	30	15	41.24
2	0.5	220	40	90	7	29.54
3	0.5	220	60	30	7	41.28
4	0.5	220	60	90	15	29.44
5	0.5	270	40	30	7	37.07
6	0.5	270	40	90	15	26.62
7	0.5	270	60	30	15	35.77
8	0.5	270	60	90	7	41.18
9	20.0	220	40	30	7	30.12
10	20.0	220	40	90	15	19.81
11	20.0	220	60	30	15	30.10
12	20.0	220	60	90	7	19.81
13	20.0	270	40	30	15	25.92
14	20.0	270	40	90	7	15.12
15	20.0	270	60	30	7	25.96
16	20.0	270	60	90	15	15.10
17	0.5	245	50	60	11	31.74
18	20.0	245	50	60	11	21.66
19	10.0	220	50	60	11	28.15
20	10.0	270	50	60	11	22.43
21	10.0	245	40	60	11	24.27
22	10.0	245	60	60	11	24.27
23	10.0	245	50	30	11	32.81
24	10.0	245	50	90	11	20.25
25	10.0	245	50	60	7	24.28
26	10.0	245	50	60	15	24.25
27	10.0	245	50	60	11	24.27

Table 8.3 Variance analysis of DOE1experiment

Parameter	SS	df	MS	F_0	P-value
(1) Injection time (L)	675.6	1	675.6	233.71	0.000005
Injection time (Q)	13.5	1	13.5	4.68	0.073814
(2) Injection temperature (L)	33.0	1	33.0	11.42	0.014880
Injection temperature (Q)	1.5	1	1.5	0.52	**0.497633**
(3) Mold temperature (L)	9.6	1	9.6	3.33	0.117968
Mold temperature (Q)	0.1	1	0.1	0.05	**0.835974**
(4) Packing pressure (L)	386.8	1	386.8	133.80	0.000025
Packing pressure (Q)	10.0	1	10.0	3.47	0.111928
(5) Packing Time (L)	14.3	1	14.3	4.96	0.067622
Packing time (Q)	0.1	1	0.1	0.05	**0.832551**
Interaction between 1L and 2L	17.6	1	17.6	6.10	0.048428
Interaction between 1L and 3L	10.8	1	10.8	3.75	0.101108
Interaction between 1L and 4L	11.6	1	11.6	4.00	0.092565
Interaction between 1L and 5L	15.7	1	15.7	5.45	0.058317
Interaction between 2L and 3L	11.2	1	11.2	3.86	0.097107
Interaction between 2L and 4L	19.0	1	19.0	6.58	0.042664
Interaction between 2L and 5L	15.5	1	15.5	5.37	0.059674
Interaction between 3L and 4L	15.4	1	15.4	5.32	0.060644
Interaction between 3L and 5L	23.9	1	23.9	8.27	0.028206
Interaction between 4L and 5L	11.0	1	11.0	3.81	0.098703
Error	17.3	6	2.9		
Total	1389.5	26			

The response surface shown in Fig. 8.6 can be described as a second-order polynomial, as shown in Eq. 8.1. Table 8.5 shows equation coefficients.

$$\begin{aligned} \mathrm{TRS} = {} & a + b\ \mathrm{T_{inj}} + c\ \mathrm{T_{inj}^2} - d\ \mathrm{P_{rec}} + e\ \mathrm{P_{rec}^2} - 245f\ \mathrm{T_{inj}} - 50g\ \mathrm{T_{inj}} - h\ \mathrm{T_{inj}}\ \mathrm{P_{rec}} \\ & + 11i\ \mathrm{T_{inj}} + 245j\ \mathrm{P_{rec}} + 50l\ \mathrm{P_{rec}} - 11m\ \mathrm{P_{rec}} - n \end{aligned} \tag{8.1}$$

where:

TRS Surface Residual Stresses [MPa];
$\mathrm{T_{inj}}$ injection time [s];
$\mathrm{P_{rec}}$ packing pressure [MPa];

In order to model the response surface of Fig. 8.6, the following were considered constant: Injection Temperature 245 °C, Mold Temperature 50 °C, and Packing Time equal to 11 s.

Validation of results was carried out through normality analysis and residual randomness (mean deviation) of the dependent variable, that is, Surface Residual Stresses. The plotting graph in Fig. 8.7 confirms the hypothesis of normal distribution of the dependent variable, where residue values are on or close to the line

Table 8.4 Variance analysis of DOE1 experiment, disregarding the variables injection temperature (Q), mold temperature (Q) and packing time (Q)

Parameter	SS	df	MS	F_0	P-value
(1) Injection time (L)	675.6	1	675.6	322.07	0.00000002
Injection time (Q)	20.4	1	20.4	9.74	0.01230733
(2) Injection temperature (L)	33.0	1	33.0	15.73	0.00327234
(3) Mold temperature (L)	9.6	1	9.6	4.58	0.06090199
(4) Packing pressure (L)	386.8	1	386.8	184.39	0.00000027
Packing pressure (Q)	15.4	1	15.4	7.34	0.02405852
(5) Packing time (L)	14.3	1	14.3	6.83	0.02811636
Interaction between 1L and 2L	17.6	1	17.6	8.41	0.01759242
Interaction between 1L and 3L	10.8	1	10.8	5.16	0.04921753
Interaction between 1L and 4L	11.6	1	11.6	5.51	0.04354916
Interaction between 1L and 5L	15.7	1	15.7	7.51	0.02284851
Interaction between 2L and 3L	11.2	1	11.2	5.32	0.04654042
Interaction between 2L and 4L	19.0	1	19.0	9.06	0.01471052
Interaction between 2L and 5L	15.5	1	15.5	7.40	0.02359824
Interaction between 3L and 4L	15.4	1	15.4	7.32	0.02413782
Interaction between 3L and 5L	23.9	1	23.9	11.40	0.00817684
Interaction between 4L and 5L	11.0	1	11.0	5.25	0.04760392
Error	18.9	9	2.1		
Total	1389.5	26			

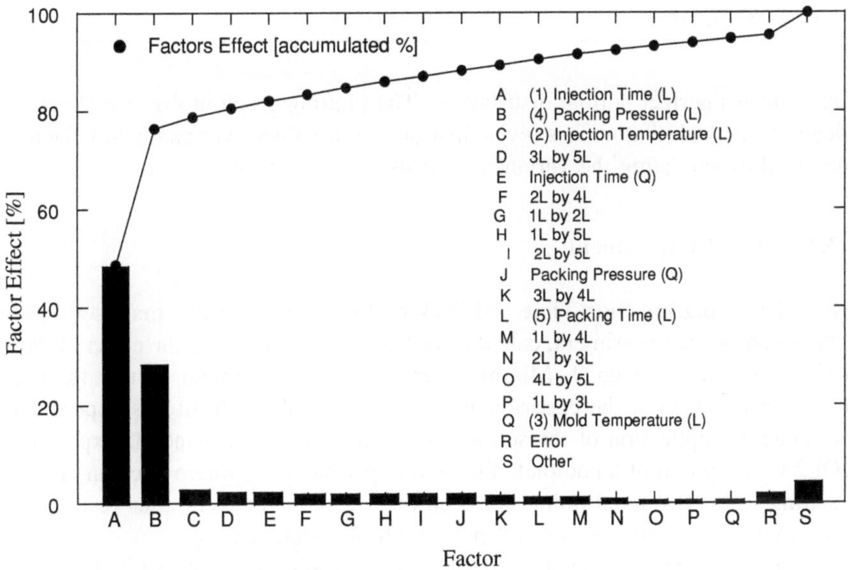

Fig 8.5 Pareto chart, DOE1 experiment

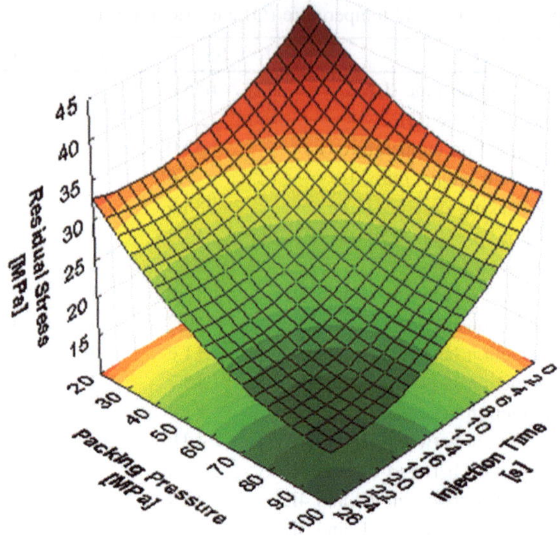

Fig. 8.6 Response surface of DOE1 experiment

Table 8.5 Equation 8.1 coefficients

a = 73.57349387749	f = 0.004308049671	l = 0.0032666666667
b = 0.209157973303	g = 0.008436815193	m = 0.006916666667
c = 0.026060638776	h = 0,002904821037	n = 23.95336
d = 0.854556273841	i = 0.0254386413447	
e = 0.002387407407	j = 0.0014533333334	

that indicates perfect normal distribution. The plotting graph in Fig. 8.8 shows an adequate residual randomness (even distribution of values over and below the axis $x = 0$), thus validating the variance analysis previously carried out.

8.3.2.2 DOE2 Experiment

After identifying Injection Time and Packing Pressure as the two main settings, a central composite experiment, called DOE2, aiming at analyzing the effect of these two settings was designed. Minimum, middle, and maximum values for each setting (parameter) of the experiment are given in Table 8.6. In this experiment, the range of application of the setting Injection Time was expanded. Experiment DOE2 is composed of a complete factorial experiment 22, where one can analyze 4 settings, plus two axial points for each one of the parameters, thus resulting in four other settings and one central point with one more setting.

Therefore, experiment DOE2 has a total of nine settings, as given in Table 8.7, where response variable is Surface Residual Stress values obtained through simulation of the injection molding process carried out through the Moldflow software. Values are shown in the last column.

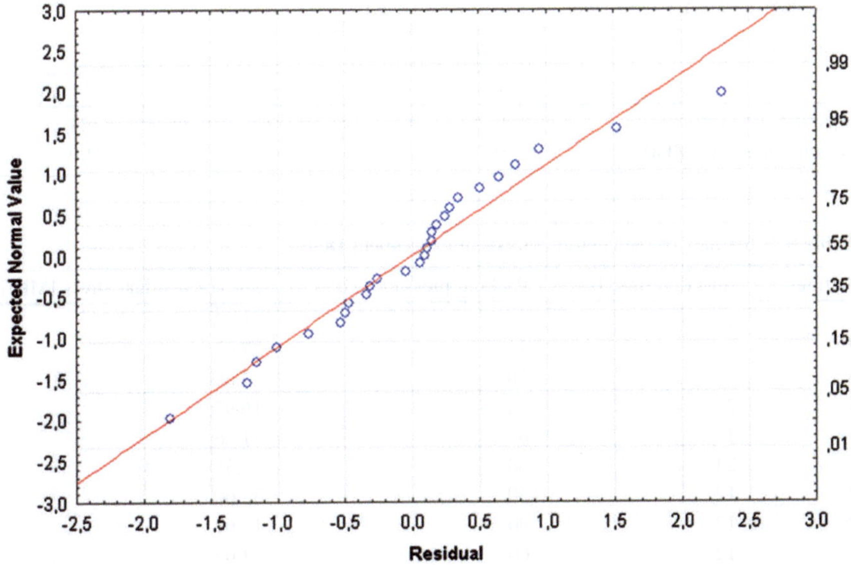

Fig. 8.7 Normal probability plotting chart

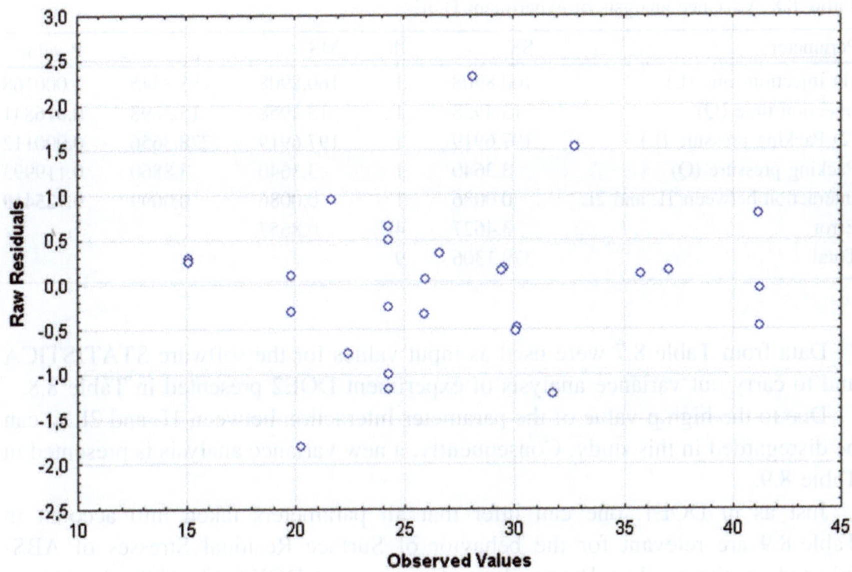

Fig. 8.8 Residual randomness plotting graph

Table 8.6 Parameters and levels of experiment DOE1

Parameter	Level		
	Minimum	Middle	Maximum
Injection time (s)	1	12	24
Packing pressure (MPa)	30	60	90

Table 8.7 Settings of the central composite experiment DOE2

Settings	Injection time [s]	Packing pressure [MPa]	Surface residual stress [MPa]
1	1	30	38.42
2	1	90	27.63
3	24	30	27.60
4	24	90	16.93
5	1	60	31.86
6	24	60	22.31
7	12	30	32.60
8	12	90	19.61
9	12	60	24.05

Table 8.8 Variance analysis of experiment DOE2

Parameter	SS	df	MS	F_0	P-value
(1) Injection time (L)	160.8908	1	160.8908	185.8545	0.000168
Injection time (Q)	13.4958	1	13.4958	15.5898	0.016841
(2) Packing pressure (L)	197.6919	1	197.6919	228.3656	0.000112
Packing pressure (Q)	3.3640	1	3.3640	3.8860	0.119993
Interaction between 1L and 2L	0.0086	1	0.0086	0.0099	**0.925410**
Error	3.4627	4	0.8657		
Total	379.1306	9			

Data from Table 8.7 were used as input values for the software STATISTICA and to carry out variance analysis of experiment DOE2 presented in Table 8.8.

Due to the high p-value of the parameter Interaction between 1L and 2L, it can be disregarded in this study. Consequently, a new variance analysis is presented in Table 8.9.

Just as in DOE1, one can infer that all parameters taken into account in Table 8.9 are relevant for the behavior of Surface Residual Stresses of ABS-injected specimens. In a Pareto chart of experiment DOE2 (Fig. 8.9), the parameter Injection Time is shown to have a higher relevance than the Packing Pressure parameter. Together, they account for 95 % of variance in analysis of Surface Residual Stresses of experiment DOE2.

The response surface that correlates Surface Residual Stresses with parameters Injection Time and Packing Pressure obtained through the STATISTICA software

Table 8.9 Variance analysis do experiment DOE2 disregarding the parameter Interaction between 1L and 2L

Parameter	SS	df	MS	F_0	P-value
(1) Injection time (L)	160.8908	1	160.8908	231.7427	0.000022
Injection time (Q)	13.4958	1	13.4958	19.4390	0.006963
(2) Packing pressure (L)	197.8004	1	197.8004	284.9063	0.000013
Packing pressure (Q)	3.3640	1	3.3640	4.8454	0.078971
Error	3.4713	5	0.6943		
Total	379.1306	9			

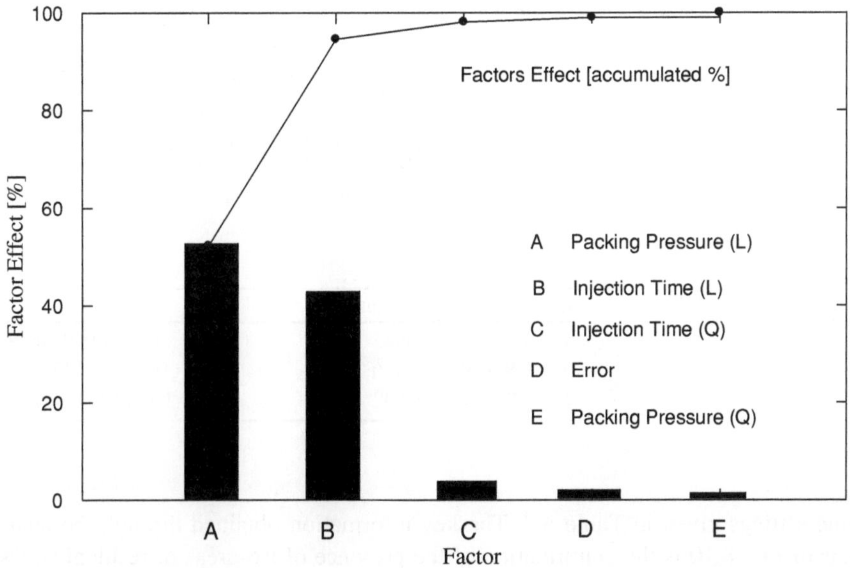

Fig. 8.9 Pareto chart, experiment DOE2

is represented by Fig. 8.10. Equation (8.2) represents the second-order model describing such response surface. Equation coefficients are given in Table 8.10.

$$\mathrm{TRS} = a - b\ \mathrm{T_{inj}} + c\ \mathrm{T_{inj}^2} - d\ \mathrm{P_{rec}} + e\ \mathrm{P_{rec}^2} + f\ \mathrm{T_{inj}}\ \mathrm{P_{rec}} \tag{8.2}$$

where:

TRS Surface Residual Stresses [MPa];
$\mathrm{T_{inj}}$ Injection Time [s];
$\mathrm{P_{rec}}$ Packing Pressure [MPa];

A response surface of experiment DOE2 in Fig. 8.10 is a graphic demonstration of the solution to Eq. 8.2. It also shows the position of residual stress values of the

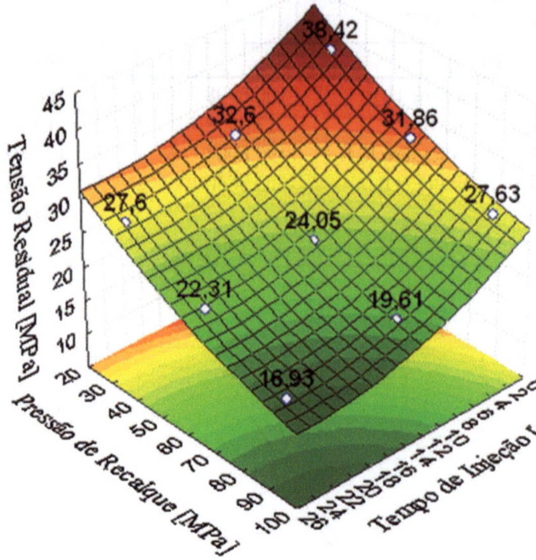

Fig. 8.10 Response surface do experiment DOE2

Table 8.10 Equation 8.2 coefficients below

Equation 8.2 coefficients	
a = 49.109512013923	d = 0.3531409979609
b = 0.9140054624976	e = 0.0013341269841
c = 0.0182262061609	f = 0.00013434089001

nine settings given in Table 8.7. The key information obtained through the analysis of Fig. 8.10 is the confirmation of the presence of iso-areas of residual stress values, previously demonstrated by experiment DOE1. Statistical validation of experiment DOE2 was carried out through the analysis of the normal probability chart (Fig. 8.11), which shows an adequate behavior in relation to the normal ideal distribution.

8.3.2.3 First Trial

Upon the response surface of Fig. 8.10, the points corresponding to experiments 1 and 2 which were obtained through the injection molding process, and were subjected to chemical attack were positioned. Respective settings can be seen in Tables 8.11 and 8.13, leading thus to Fig. 8.12, further discussed.

Table 8.11 shows the main parameters used for the injection of specimens. The main characteristic is that packing pressure is lower than injection pressure, which led to different simulated residual stress in the trials.

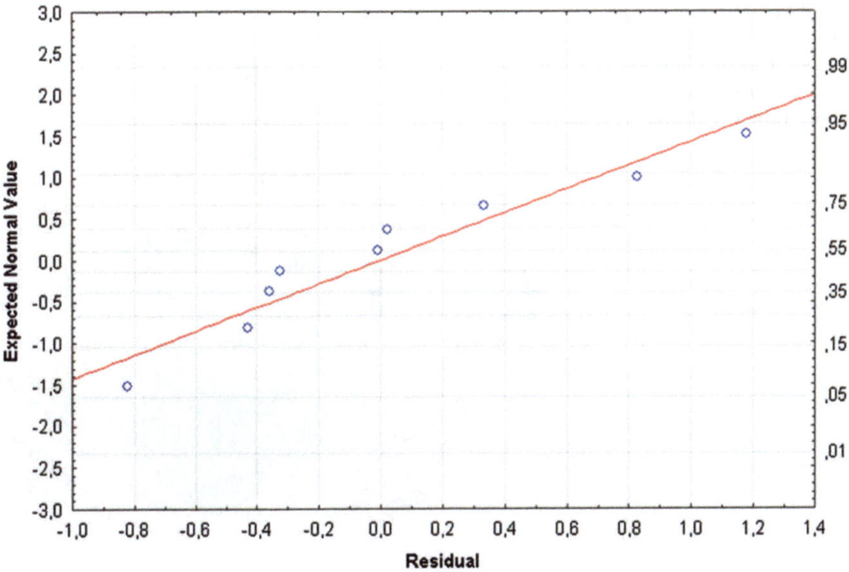

Fig. 8.11 Normal probability chart, experiment DOE2

Table 8.11 Main process settings used for injecting specimens

Parameter	Settings			
	1	6	12	24
Injection time (s)	1	6	12	24
Injection temperature (°C)	240	240	240	240
Mold temperature(°C)	52	49	48	51
Packing pressure (MPa)	47.2	35.8	37.4	39
Packing time (s)	10	10	10	10
Residual stress (MPa)	34.5	33	29.5	26

Figure 8.13 shows the photographs obtained through electron microscopy of the four trials described in Table 8.11. In these cases, as packing pressure is lower than injection pressure, one can observe that for low injection times the surface was more severely attacked (injection times 1 and 6 s). Part bleaching was observed upon visual inspection. For higher injection times, 12 and 24 s, specimen surface showed no bleaching or acid attack, therefore no micro cavities were present at this scale of observation. Through the similarity of simulated and parameters obtained experimentally, the four results on response surface of Fig. 8.12 were highlighted. These are indicated by points 1, 6, 12, and 24.

Figure 8.12 also shows the presence of two cases separated by an isoline, those which were attacked by the acid and those which were not. Therefore, the hypothesis is that by following an isoline corresponding to point 24, one can obtain

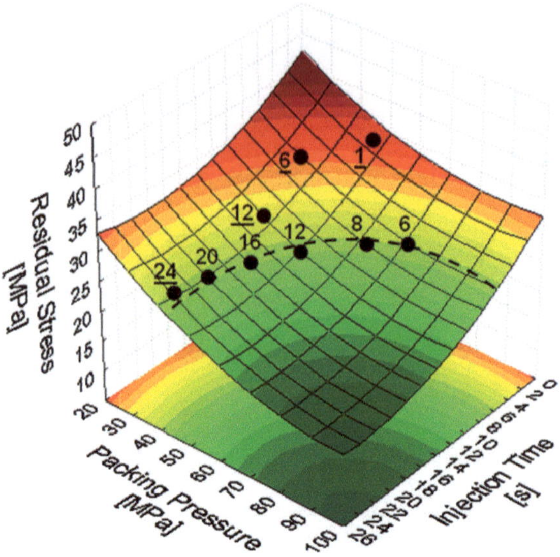

Fig. 8.12 Response surface obtained using cenítral composite design, results obtained through chemical probe. Tables 8.11 and 8.13

parts which would not be attacked by the acid through low injection times and high packing pressure. Besides, no condition below this isoline should present bleaching when attacked with acetic acid.

8.3.2.4 Second Trial

Table 8.12 was designed as follows: Eq. 8.2 was solved taking into account constant residual stress and equal to 26.05 MPa, a condition required to meet the hypothesis described above.

The same injection times, molding and injection temperatures from Table 8.13 were used. Packing pressure (equation) was thus obtained. Upon entry of the same data into the flow rate simulation tool, simulated packing pressure and post-processed residual stresses are determined.

The highest error between simulated results and those obtained by Eq. (8.2) for packing pressure was 3.6 %, which corresponds to point 20. Five samples analyzed, corresponding to points 6, 8, 12, 16, and 20 showed no bleaching after treatment with acid. Electron microscopy analysis with 200 magnification showed no presence of micro cavities in the order of magnitude observed in the first trial (see Fig. 8.14). Process settings of trials of response time in Fig. 8.12 are transferred via analogy. Taking into account the absence of surface bleaching of this trial, one can state that, in these conditions, all parts would be adequate for the chrome plating process. Consequently, the hypothesis of following the isoline would be demonstrated, i.e., it

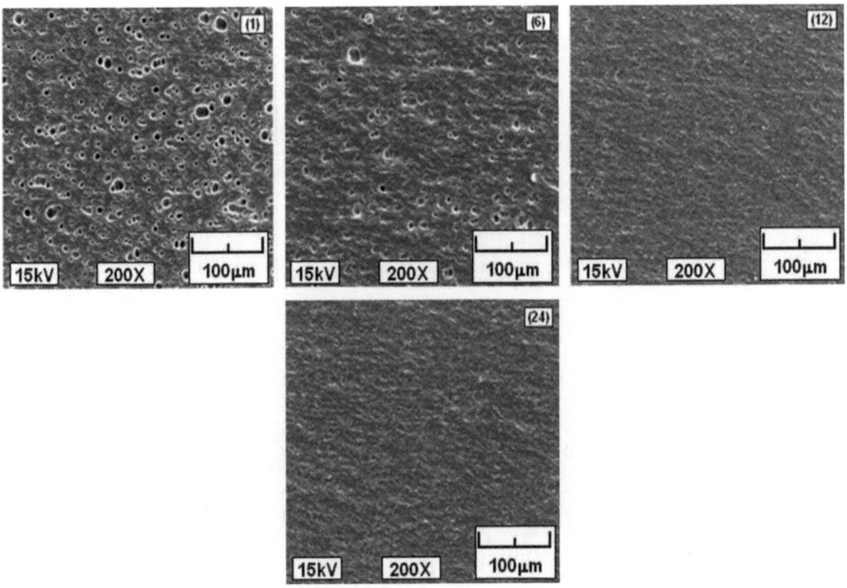

Fig. 8.13 Scanning electron microscopy magnified 200 times, settings 1, 6, 12 and 24, after immersion in acetic acid

Table 8.12 Solution to Eq. 8.2 for constant residual stress

Residual stress (MPa)	Injection time (s)	Packing pressure (MPa)
26.05	6	70.64
26.05	8	63.16
26.05	12	52.21
26.05	16	45.04
26.05	20	40.77

Table 8.13 Main process settings used for injecting specimens with constant surface residual stress

Parameter	Settings				
	6	8	12	16	20
Injection time (s)	6	8	12	16	20
Injection temperature (°C)	240	240	240	240	240
Mold temperature(°C)	50	50	51	51	51
Packing pressure (simulated) (MPa)	71.6	63.5	53.7	45.6	42.3
Packing time(s)	15	15	15	15	15
Residual stress (simulated) (MPa)	25.9	26	25.7	25.9	25.7
Solution of the Eq. 8.2 to constant residual stress of 26,05 MPa					
Packing pressure (equation) (MPa)	70.64	63.16	52.21	45.04	40.77

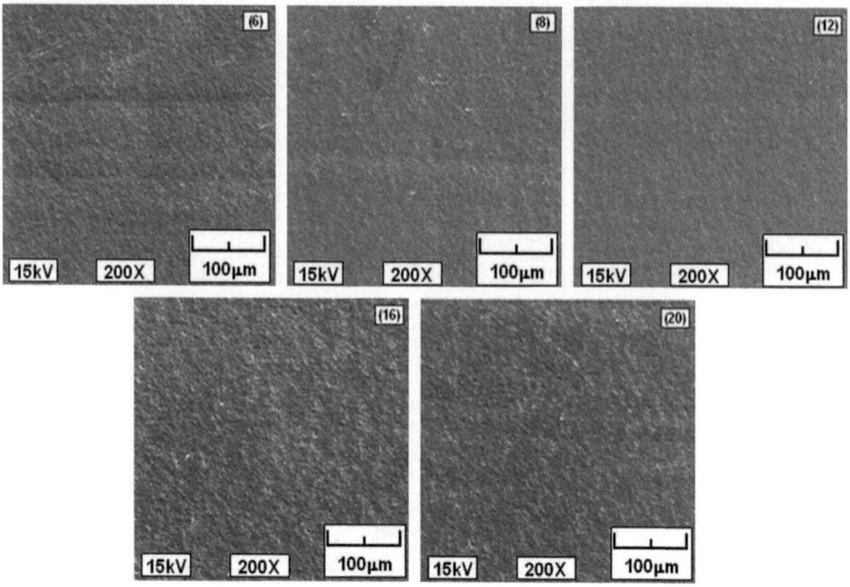

Fig. 8.14 Scanning electron microscopy magnified 200 times, settings 6, 8, 12, 16 and 20, after immersion in acetic acid

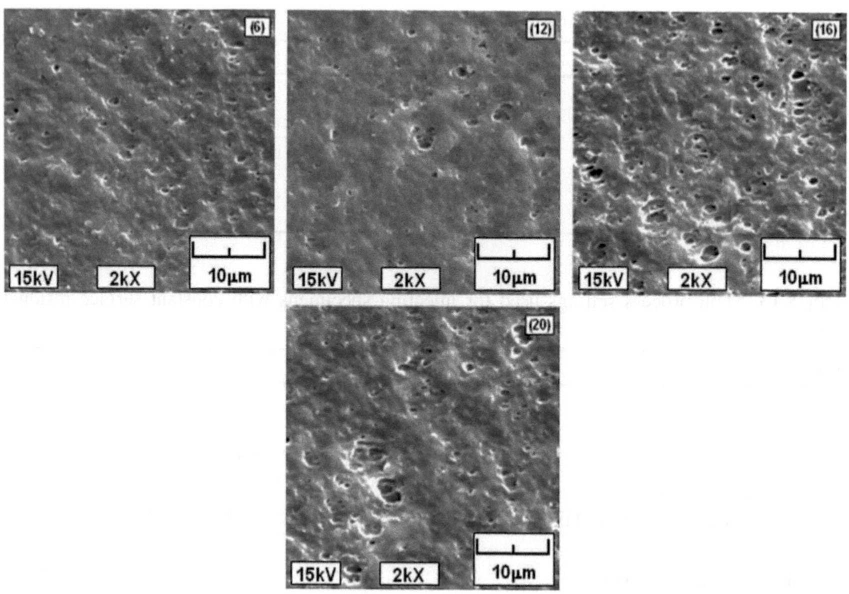

Fig. 8.15 Scanning electron microscopy magnified 2,000 times, settings 6, 8, 12, 16 and 20, after immersion in acetic acid

is possible to obtain injected parts with lower injection times and packing pressure equal to injection pressure. Figure 8.15 shows the images from Fig. 8.14 magnified 2,000 times. At this magnitude, micro cavities on the surface can be observed. However, micro cavity size is shown to be reduced as injection time decreases. In the least favorable condition, the order of magnitude is 3 μm at most. Literature shows that a well-defined rugosity in terms of Ra and Rz is beneficial to the quality of polymer/metallization interface bonding [11].

8.4 Conclusion

Analysis of thickness of layers deposited has shown that these are in the range recommended by the literature. Optical microscopy with image capture has shown to be adequate for this analysis. Copper layer could be decreased in 1 μm and nickel layer could be decreased in 5 μm. The method used for analyzing the morphology of the treated surface correlated to the surface residual stress obtained through ABS injected molding has shown to be consistent with the use of a combination of statistical, numerical, and experimental tools. The tool for statistical design of trials has helped in organizing and defining the setting parameters of injection to be tested. This in turn has enabled greater efficiency in the relationship between number of trials and data on the relevance of each parameter. The central composite design statistical experiment has shown to be the most adequate for achieving response surfaces in the trial analysis phase. The major contribution from this study is the demonstration through trials that it is possible to obtain surfaces with adequate residual stress which are not significantly affected by acetic acid, therefore being adequate surfaces for the metallization process. Parts were obtained from low to high injection times where packing pressure was always equal to injection pressure. A more detailed analysis through electron microscopy with 2000 magnification showed that, for lower injection times with roughly the same simulated residual stress, the surface is less susceptible to acid attack.

References

1. Murakamik, H.E., Bom, R.P.: Process parameters influence on residual stress on injected parts surface. In: Proceedings of the 63rd ABM Annual Congress, Santos, p. 626–635 (2008)
2. Murakami, E.H.: (2009) Process parameters influence over the ABS injected parts residual surface tension. Master Science Thesis on Materials Science and Engineering, University of Santa Catarina State, CCT, Joinville, p. 99
3. Turnbull, A., Maxwell, A.S., Pillai, S.: Residual stress in polymers—evaluation of measurement techniques. J. Mater. Sci. **34**, 451–459 (1999)
4. Jansen, J.A.: Environmental stress cracking—examples from the automotive industry, Society of Plastics Engineers, ANTEC. In: Proceedings of Plastics: Annual Technical Conference, vol. 3, Special Areas (2005)

5. Ryntz, R.A., Yaneff, P.: Coating of Polymers and Plastics. Marcel Dekker Inc, New York (2003)
6. Hindle, C.S., White, J.R., Dawson, D., Thomas, K.: Polym. Eng. Sci. **32**, 157 (1992)
7. Turnbull, A., Maxwell, A.S.: Measurement of residual stress in engineering plastics using the hole-drilling technique. Polym Test Elsevier **22**(2), 231–233 (2003)
8. Maxwell, A.S., Turnbull, A.: Characterisation of environment stress cracking susceptibility of weldline defects by the chemical probe technique. National Physical Laboratoy REPORT CMMT(A)269 (2000)
9. Chang, R.Y., Chang, Y.R., Peng, Y.H., Yang, W.H., Hsu, C.H.: True 3d numerical method of flow-induced Residual stress prediction in injection molding. In: Proceedings of The Polymer Processing Society 23rd, Annual Meeting—PPS-23. Salvador, BA, Brazil (2007)
10. Multinal group http://www.multinal.com/metallizing-of-abs-synthethics-products-spareparts-chrome-nickel-plating.html. Accessed 09 Jun 2014
11. Kurek, A.P., Sellin, N., Geelsleichter, M.: More clean technology applied on ABS conditioning step for metallization. In: Proceedings of the 8th CBECiMat—Brazilian Materials Engineering and Science Congress, Porto de Galinhas, PE, Brazil, p. 10730–10741 (2008)
12. ASTM D 1939–84 Practice for determining residual stresses in extruded or molded acrylonitrile-butadiene-styrene (ABS) parts by immersion in glacial acetic acid (Withdrawn 1999). http://www.astm.org/Standards/D1939.htm. Accessed 09 Jun 2014
13. Autodesk Moldflow Insight. Release 2012, educational version. http://www.autodesk.com/products/autodesk-simulation-family/features/simulation-moldflow/all/list-view. Accessed 09 Jun 2014
14. Statistica Software. Release 9, educational version. http://www.statsoft.com/Resources/Support/Software-Updates. Accessed 09 Jun 2014

Chapter 9
Residual Stress Evaluation in Vitreous Enameled Steel Sheets by Digital Images Analysis of Microstructures

Rita Ambu, Andrea Zucchelli and Giangiacomo Minak

Abstract Composite vitreous-enameled steel sheet have become more attractive for structural applications. In this chapter, an investigation on the residual stresses, introduced during the manufacturing of these materials, is reported. The local structure evidences the presence of bubbles in the coating and the interface domain between the substrate and the external layer made of a complex material system where the ceramic and the metal constituents are mixed. This particular local structure can affect the distribution and entity of the residual stresses. Digital images of cross-section micrographs have been processed and used to discretize the material microstructure into a mesh suitable for finite element analysis. The residual stresses have been evaluated on coatings manufactured with different technologies and, in particular, the effect of the presence of bubbles on the stresses has been highlighted.

Keywords Vitreous enamel coating · Residual stress · Image processing · Finite element method

9.1 Introduction

Composite vitreous-enameled steel sheets are a special class of metal-ceramic composite materials obtained by a two stages coating process: the deposition of the enamel raw material over the metal surface and the firing at high temperature of

---

R. Ambu (✉)
University of Cagliari, Piazza d'Armi, 09123 Cagliari, Italy
e-mail: ambu@unica.it

A. Zucchelli · G. Minak
University of Bologna, Viale Risorgimento 2, 40136 Bologna, Italy
e-mail: a.zucchelli@unibo.it

G. Minak
e-mail: giangiacomo.minak@unibo.it

F. R. Leta (ed.), *Visual Computing*, Augmented Vision and Reality 4,
DOI: 10.1007/978-3-642-55131-4_9, © Springer-Verlag Berlin Heidelberg 2014

the metal-enamel system. A wide spectrum of industrial and domestic applications currently make use of these coatings (e.g., the treatment of components for household use, the protection of interior walls of reactors for chemical processes, the protection of mechanical components of rotary heat exchangers, and aircraft turbojets). Enameled steel components are appreciated for their esthetic properties as well as for their chemical characteristics. In particular, enameling of metal substrates is used to provide protection against chemical corrosion for applications in hostile environments such as in reactors for chemical processes. In fact, from a functional point of view, vitreous enamel coatings have an excellent resistance to chemical corrosion processes [1–3]. Steel enameling is also used to obtain low roughness surfaces, having no fouling, and/or antimicrobial properties. Vitreous enamel coatings surfaces are also characterized by high values of hardness (up to 800 HV) giving to coated components a good resistance to tribological phenomena such as abrasive wear [4]. Compared to other coatings for metals, such as thermally sprayed ceramics, vitreous enamel coatings are characterized by a chemical and not only a physical adhesion to the substrate achieved by a graded interface that is developed during the coating firing process. Enameled steel components are also subjected to internal stresses. From a macroscopic point of view, these internal stresses contribute to prevent the instability of enameled steel plate when subjected to impact event [5]. But, the presence of residual stresses also affects the coating failure due to cracks onset and propagation, the spallation of the coating, the shape changes of the coated components and in general can influence the performance of the entire coated parts. Even if enameling is nowadays an industrial practice, some of the mechanical performances of enameled composites have not been studied and others are not completely understood. Based on this fact, the present work is intended to add an original contribution about one of the main physical and microstructural aspects that characterize the enamel coating: the residual stress and its relationship with the local characteristics of the enamel matrix.

Residual stresses are introduced in the enameled substrate during the manufacturing process. In fact, during the cooling to room temperature thermal mismatch stress develops due to the difference between the coefficients of thermal expansion (CTE) of the vitreous enamel material and the metal substrate. Different methods have been used to measure the residual stresses in coatings: mathematical modeling [6, 7], mechanical methods [8], and material removal techniques [9]. Each technique has certain advantages and limitations; their applicability is determined by such factors as shape, dimensions, materials of the coating and the substrate, knowledge of the constituents' properties and processing conditions. The experimental methods can be either destructive or non destructive, but all use a global approach to the evaluation of the residual stress. The computational models may be better tools to identify potential problems and decide what changes in the coating process are necessary to control the residual stresses. However, most models of residual stresses assume that coatings are homogeneous in structure and therefore cannot predict stress concentrations. But, for a more exhaustive mechanical characterization, the study of the local distribution of the residual stress is an important task. In particular, the identification of regions of high stress level in the coating structure is of major concern. In fact, the configuration of the local structure of the material, such as the presence of bubbles or the existence of a transition zone between the two constituents of the material, can affect the local values of the stresses, thus promoting or delaying the onset of the fracture of the material in service. This can be accomplished by directly evaluating the residual stress on the real microstructure of the material, obtained from digital micrographs processed to this task. Digital image processing, widely used in many fields of sciences and engineering, has been applied in material sciences for quantitative description of complex material microstructures [10, 11] and, in particular, for stress investigation in thermal spray coatings [12, 13].

In this chapter, a procedure based on the analysis of digital images of enamel coatings has been applied to the measurement of thermal stresses. The approach considered uses the digital images filtered and converted into a graphical format to perform finite element simulations on the real local structure of the material. In this study, coatings manufactured with different technologies were considered. A preliminary characterization of the microstructures was performed by quantitative measurements on digital images. Then, the thermal stresses were evaluated taking into account F.E.M. discretization of the coating structure of growing complexity. The results of the local approach were finally related to the characteristics of the microstructure.

9.2 Enamel Coating and Enameling Technology

Vitreous enamels are inorganic materials that are used as coatings for metallic components. These materials are based on a special ceramic-vitreous matrix in which specific additives are randomly dispersed. The ceramic-vitreous matrix is made by a mixture of various row materials and elements and in particular it is based on boron-silicate glass added with metal oxides of titanium, zinc, tin, zirconia, alumina, etc. These additions are often used to enhance some important performances such as corrosion and wear resistance, mechanical strength, fracture toughness, and also esthetic functions. The compound obtained by mixing these rough materials is then fired at about 1,300 °C and quickly cooled into cold water, as shown in the scheme of Fig. 9.1. At the end of this process, the ceramic-vitreous matrix is obtained and it is commonly called frit.

In order to obtain the vitreous enamel raw materials, specific additives have to be mixed and milled with the frits. These additives are necessary to increase the adherence of the enamel to the metal substrate and also to give special properties and functions to the final coat. The mixing-milling process can be realized by a wet or a dry process. In the wet process, the vitreous enamel raw material consists on a slip in which the small grain of the frits and of the additives are homogeneously dispersed in water. In the case of the dry milling process, the addition of silicone oil enables the production of fine powder made by frits and additives

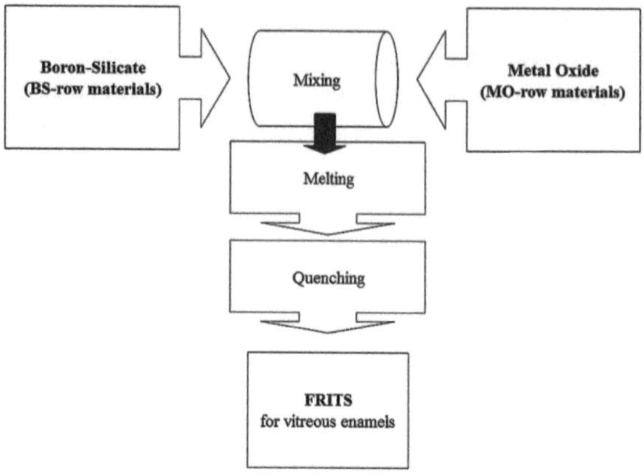

Fig. 9.1 Scheme of the frits for vitreous enamel raw material production

grains that can be applied to the metal substrate by an electrostatic deposition technology.

The wet enameling process consists of four phases shown in Fig. 9.2: (i) wet milling, (ii) wet spraying (or flow coating deposition), (iii) drying, and (iv) firing at high temperature (about 550 °C for low melting alloys, about 750 °C for cast iron and about 850 °C for steel). The dry process is based on three phases: (i) dry milling with silicone oil addition, (ii) electrostatic deposition, and (iii) firing at high temperature (about 550 °C for low melting alloys, about 750 °C in case of cast iron and about 850 °C for steel).

Figure 9.3 gives an example of the firing curve. This figure shows the typical firing diagram where the heating, the maturing, and the cooling phases are highlighted together with some photos of the enamel pellets over a metal substrate taken during the firing process. The photos of the pellets, taken with a heated microspore, show the enamel pellet transition from A up to B temperature during the firing phase, and the complete fusion of the enamel pellet during the maturing phase (C). It is interesting to note that during the cooling phase the enamel lens (D and E) does not change its shape and it consolidates over the metal surface.

During the firing process, the enamel raw material melts and interacts with the metal substrate, thus enabling the formation of a continuous varying structure.

As shown in the micrograph of the transversal section in Fig. 9.4, the interface zone between the substrate and the external layer is made of a complex material system where the vitreous enamel and the metal constituents are mixed. In particular, three main regions can be identified, starting from the bottom of Fig. 9.4a: a first region is made of metal, the second region is the interphase where both metal and enamel constituents are mixed, and the third region is composed by the vitreous enamel material. Referring to Fig. 9.4b, the presence of metallic dendrites that hinder the substrate and the external layer passing through the interphase region can also be noted.

As already mentioned in the introduction, the mechanical behavior of the vitreous enamel-metal composite at room temperature is greatly influenced by

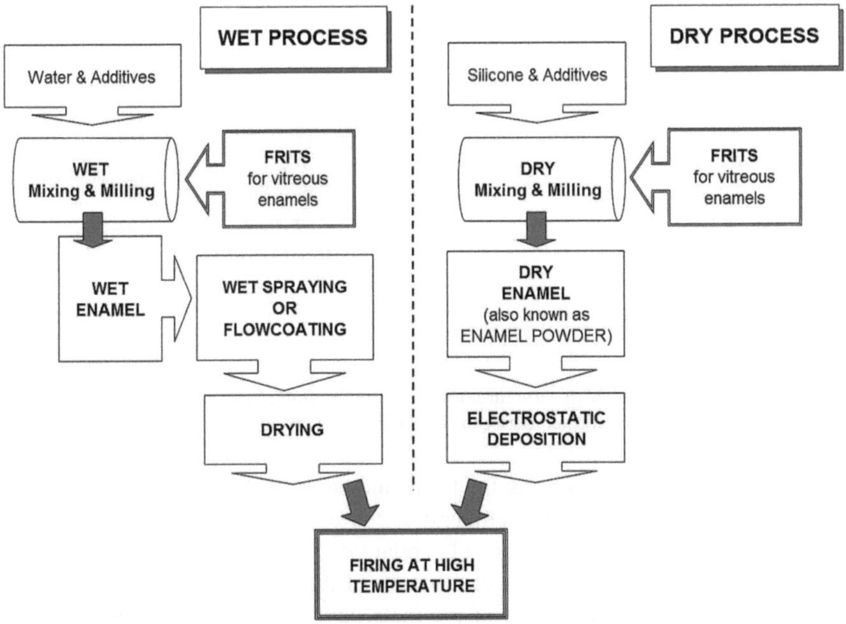

Fig. 9.2 Scheme of the enameling coating processes in the case of wet enamel or dry enamel deposition techniques

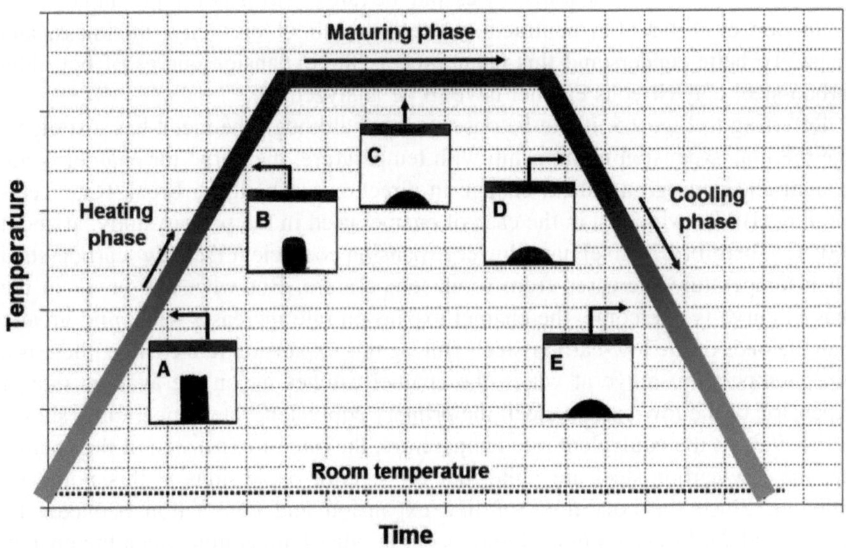

Fig. 9.3 Scheme of firing curve for the enamel-substrate system with details of the enamel transition during the thermal treatment

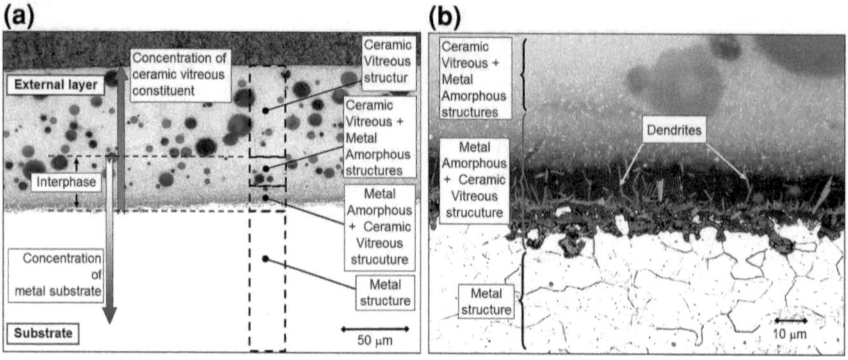

Fig. 9.4 Enamel-steel composite structure. **a** General micrograph view of the enamel and metal composite; **b** detail of the interface between the enamel coating and the metal substrate

internal stresses developed during the firing and cooling phases. In the firing of an enamel, the sealing at elevated temperatures is accomplished without excessive stress development because the enamel is relatively fluid at the firing temperatures and can easily assume the surface dimensions of the metal to which it is applied. Any difference in the contraction of the metal and the enamel is of little consequence until the temperature is reached at which the cooling rate is too rapid to allow flow of one on the other or internal flow in either. The stress initially developed is only a fraction of that which would be expected when considering the difference in the rates of contraction of metal and enamel. However, as cooling proceeds, the amount of flow decreases and the rate of stress development becomes a function of differential contraction rates, moduli of elasticity, thickness, and shape. To better understand this phenomenon, the expansion curves of both low carbon-steel and vitreous enamel have to be analyzed.

Referring to Fig. 9.5, it can be observed that although the metal has a straight-line thermal expansion relationship with temperature, the curve for enamel is not straight and undergo a radical change in direction at about the Break Rate Temperature (BRT), which is, in the case of enamel used in the present study, at about 400 °C. Therefore, enamel has a lower expansion coefficient than low-carbon-steel when temperature increases from room temperature. But when the break in the enamel curve is overcome, the enamel expansion rate increases becoming greater than the ones of the low-carbon-steel. During this expansion rate change, there is a small temperature range at which the enamel catches up on the average rate of expansion of the low-carbon-steel: the primary equivalent rate temperature, shown in Fig. 9.3. At the equivalent rate temperature, the rate of expansion of the enamel and the low-carbon-steel are alike, but the overall expansions at this point are different. Since the conditions of like expansion and contraction between the enamel and the low-carbon-steel may occur at other temperature when the enamel experiences crystallization, contains crystals which invert, or is subjected to special heat treatments, these others are called secondary equivalent rate temperatures, shown in Fig. 9.5. The enamel curves show a drop at F (enamel fusion

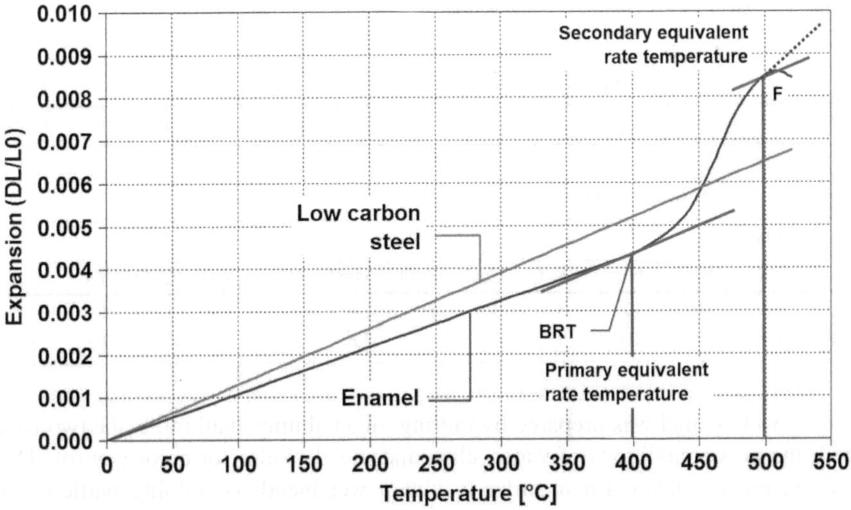

Fig. 9.5 Thermal expansion diagram of the low carbon steel and of the vitreous enamel material used in the present study

temperature) but this is a failure in the structural strength as the enamel actually continues to expand as indicated by the dotted line. Because of the high mobility of the enamel above this temperature, the stress and strain in enamel low-carbon-steel composite system become very low, even though they both continue to expand at different rates with increase in temperature. The flow of the enamel glass substantially relieves the stresses during the heating and cooling of the composite at temperatures above the BRT. The amount of tension in low-carbon-steel and compression in enamel would actually be a function of the contraction differences in cooling from BRT to room temperatures during the cooling phase. For a fixed component geometry, the intensity of residual stresses acting on the enameled steel composite can be pre-determined by means enamel design (compositions of frits and additions) aimed to match proper values of both BRT and coefficient of thermal expansion.

9.3 Materials and Characterization

Sheets of 0.8 mm thick of a very low carbon steel, as reported in Table 9.1, were coated by two blue enamels: one is a wet enamel prepared for the wet-spray application and the other one is a dry enamel for electrostatic deposition. Steel specimens have a rectangular shape: 250 mm in length and 40 mm in width.

Both enamel raw materials were prepared by the same types of frits whose compositions, given by the frits producer, are summarized in Table 9.2.

Table 9.1 Steel sheets chemical composition (%)

	Elements						
	C	Mn	Si	P	S	Al	Cu
wt%	0.003	0.31	0.03	0.025	0.04	0.025	0.032

Table 9.2 Composition of enamel frits in wt% of oxides

	Oxides											
	SiO_2	B_2O_3	Na_2O	K_2O	CaO	BaO	MnO	CoO	NiO	CuO	TiO_2	Li_2O
Frit 1	58	10	12	3	3	4	2.6	2.8	2.4	1.2	1	–
Frit 2	69	4	10	4	–	–	2.8	2.2	–	–	13	5

The wet enamel was prepared by milling, in an alumina ball mills, the two frits with the proper additions of water, clay, and metal oxides for color control. The milling process takes 4 h in order to obtain wet blends containing particles of controlled size ($\sim$45 ± 6 μm). The steel sheets for the wet enamel deposition were pre-treated according to the following seven steps: (i) degreasing at 60 °C with alkaline degreasing bath, (ii) acid attack in a 5 vol% solution of sulfuric acid at 60 °C, (iii) room temperature water bath, (iv) nickel deposition by immersion in a 1.2 % of $NiSO_4$ bath, (v) room temperature water bath, (vi) immersion in a 0.3 % neutralizer solution at 60 °C, and (vii) drying up at 110 °C.

The dry enamel was prepared by milling, in an alumina ball mills, the two frits with the proper additions of silicone oil. The milling process takes 9 h in order to obtain a dry powder with a controlled size ($\sim$28 ± 5 μm). The steel sheets used to prepare the samples by the dry enamel raw material were polished to take out all superficial impurities due to sheets machining.

After enamels deposition over the steel substrates, the all systems enamel-substrate were fired at the same conditions:

- heating phase, from room temperature to 850 °C, in 6 min;
- maturing phase, at a constant temperature of 850 °C, for 6.5 min;
- cooling phase from 850 °C to room temperature in 7 min.

In Fig. 9.6, two examples of micrographics images of the transversal section of the enameled steel are reported.

In particular, in Fig. 9.6a the transversal section of the coating obtained by means of the wet enamel is reported, and in Fig. 9.6b the coating obtained by means of the dry enamel. From these images, it is possible to note that in the case of wet enamel the distribution of the bubbles in the coating section is almost uniform; on the contrary, in the case of the dry enamel the bubbles are dispersed near the external face of the coating. This fact is due to the raw material formulation and in particular to the presence of the clay in the wet enamel formulation. The usage of clays in the wet enamel manufacturing aims to control both the slips rheology as well as the bubbles distribution in the final coating. Moreover, in both images, it is possible to note the presence of dendrites between the coating and the substrate.

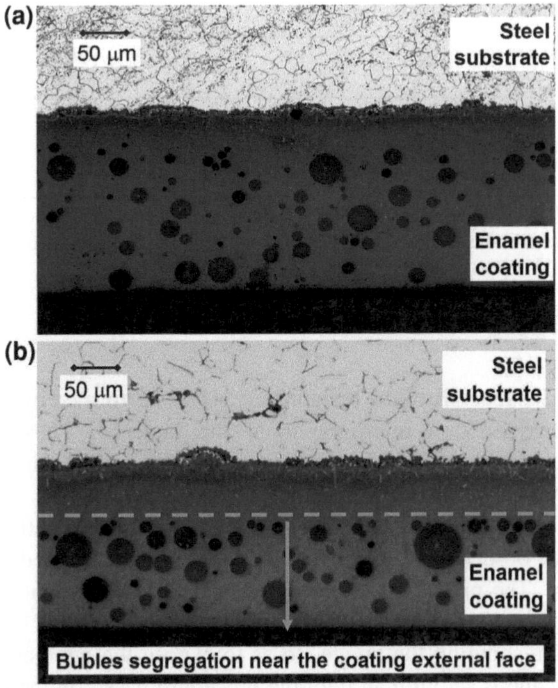

Fig. 9.6 Examples of enamel coating obtained by **a** wet enamel raw material and by **b** dry enamel raw material; in the case (**b**) the bubble segregation near the external coating face is highlighted by the *dashed line*

Table 9.3 Main characteristics of the coating microstructure

	Bubbles percentage		Bubbles diameter (μm)	
	M.V.	S.D.	M.V.	S.D.
Wet enamel	15	4	17	8
Dry enamel	17	6	15	6

In Table 9.3, the main microstructure characteristics of the enamel coatings are summarized. In particular, the bubbles density (also known as blistering) and the bubbles diameter are analyzed in terms of mean value and standard deviation.

The numerical analyses have been performed by considering the physical and the mechanical properties, respectively for the metal substrate and the enamel coating, summarized in Table 9.4.

In particular, the mechanical performance of low carbon steel was assessed by tensile tests according to UNI EN 10002. The mechanical properties of vitreous enamel material (E and ν) were estimated by four point bending tests [3]. The coefficient of thermal expansion of vitreous enamel material was determined by a heating dilatometer while the coefficient of thermal expansion for the metal was taken from the literature [1].

Table 9.4 Material properties

Material	α (10^{-5} °C^{-1})	BRT[a] (°C)	E (MPa)	ν
Low carbon steel	1.35	–	180,000	0.3
Wet enamel	1.18	400	70,000	0.27
Dry enamel	1.23	400	70,000	0.27

[a] The Break Rate Temperature (BRT), as mentioned in the Sect. 9.2, is the temperature below which, during the cooling phase, the residual stress starts developing

9.4 Methodology

The analysis of the residual stress was performed by means of the digital images of cross-section micrographs of the material. These images highlight the local characteristics of the material such as the bubbles and the graded interface between the two constituents.

In particular, the presence of bubbles in the coating is mostly evidenced, as can be observed in Fig. 9.7, where one of the analyzed images is reported.

The procedure followed for image analysis can be as summarized. A median filter (3 × 3) is first applied to the image [14, 15]. The next step consists in thresholding the image, so as to expose the geometries of interest. The threshold value is set with an iterative procedure based on the histogram image distribution. Morphological operations are then applied to the binarized image. In particular, closing was employed in order to fill in small holes, while opening was used for clustered pores.

The binary image obtained highlights the bubbles in the coating and the graded layer between the coating and the substrate as dark objects on a light background, as evidenced in Fig. 9.8.

The discretization of the coating requires the identification of the borders of the objects in the binary image. To highlight the sharp changes in intensity, an edge detection filter is applied to the image. For this task, a Sobel operator is used, based on convolving the image with a small, separable, and integer value filter in horizontal and vertical direction. The edge detection allows to produce an image where, in particular, the geometry and location of the characteristics of interest is well identified.

As known, the binary interface image uses the unit gray level to represent the interface pixels. The closed interfaces are a cluster of the pixels with the unit gray level. In order to carry out the mechanical analysis, finite element meshes must be generated for the material distribution and geometry shown in the refined binary image. The finite element meshes cannot be generated directly from the binary interface image. The discrete interface pixels have to be transformed into their geometry vector data format. This conversion is the preliminary procedure for automatic generation of the finite element meshes for the material.

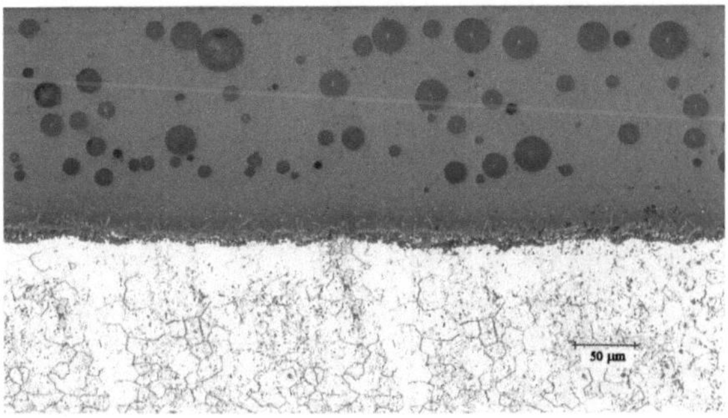

Fig. 9.7 Image of the coating

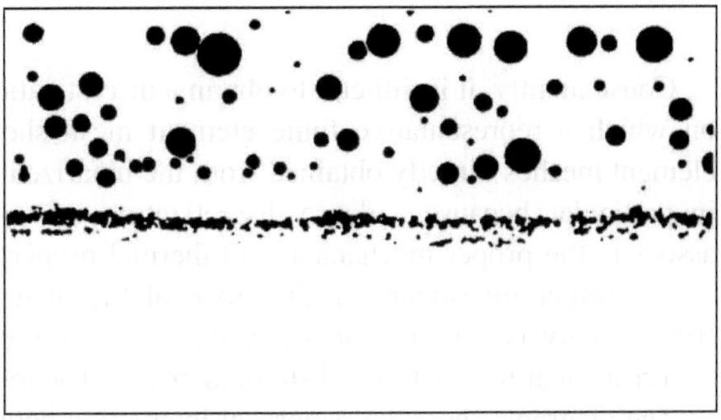

Fig. 9.8 Binarized image of the coating

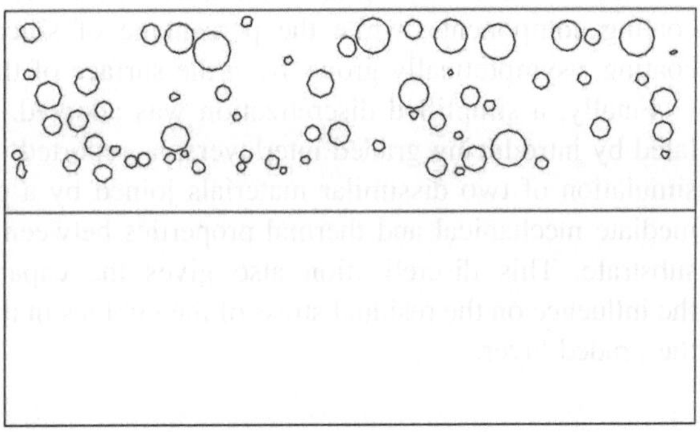

Fig. 9.9 Vectorized image of the coating

Once performed the image conversion, a scale transformation is applied to the image in the graphical format, according to the calibration data. The image magnification is 1,680 pixel/mm. The result finally obtained is reported in Fig. 9.9.

It has to be observed that in Fig. 9.9, the graded layer between the coating and the metal substrate has been disregarded.

The discretization of the zone with the graded layer between the coating and the metal substrate required a further analysis. In fact, as previously discussed, the structure of the graded layer is very complex, also if the extension can be estimated from filtered images of high magnification such as that reported in Fig. 9.10 together with a brightness diagram along with a line.

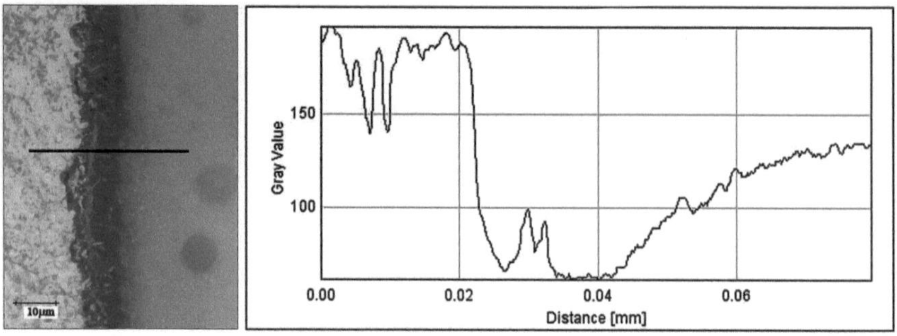

Fig. 9.10 The interface between the coating and substrate

Consequently, it is difficult to obtain a discretization of this zone of the material on which a representative finite element mesh should be generated. The finite element meshes directly obtained from the binarized images were not satisfactory, in particular because a dense discretization was obtained and was difficult to associate the proper mechanical and thermal properties in the interface zone.

To further investigate on this zone of the material, energy dispersive X-ray (EDX) analysis was used to study the composition and microstructural changes occurring near the coating-substrate interface. The results are reported in Fig. 9.11.

The diagrams show the reciprocal interaction between the coating and the steel substrate. In fact, starting from the inner substrate, steel slowly degrade into the coating components, while the percentage of silicate, basic component of the coating, asymptotically grows from the surface of the substrate.

Finally, a simplified discretization was adopted. The graded layer was simulated by introducing graded interlayers, as reported in literature for the numerical simulation of two dissimilar materials joined by a graded layer [16], with intermediate mechanical and thermal properties between those of the coating and the substrate. This discretization also gives the capability to separately evaluate the influence on the residual stress of the bubbles in the coating and the presence of the graded layer.

9.5 Results and Discussion

The vectorized images were imported in a commercial finite element code to perform the analysis of the residual stress. The finite element simulations were performed by means of ANSYS F.E.M. software. The images of the coating microstructures were discretized with quadrilateral plane elements to create a finite-element mesh adapted to fit within material boundaries. The bubbles in the microstructure were considered to be empty cells whose interstices are free to move.

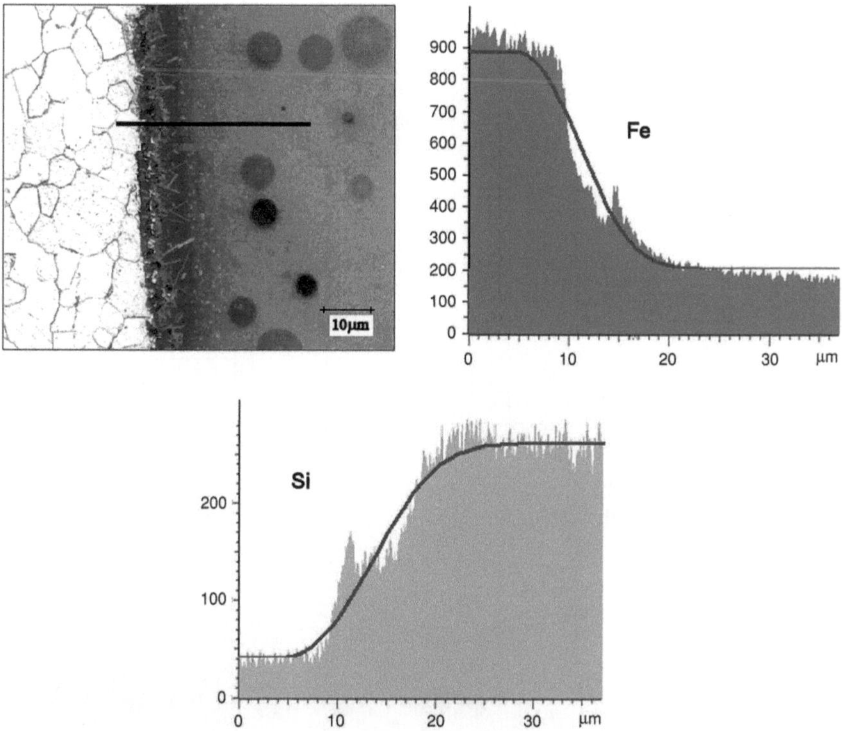

Fig. 9.11 EDX maps relative to the interface between the coating and substrate

Figure 9.12 reports a detail of the F.E.M. discretization relative to a micrograph analyzed in this study.

The analysis was performed by the simulation of the cooling of the material from the maximum temperature reached during the manufacturing to the room temperature, assuming a temperature gradient of 400 °C. In particular, micrographs of coatings representative of the two manufacturing processes before described have been considered. The thickness of the steel sheet introduced in each model is 0.8 mm. The first simulations did not took into account the presence of the graded layer, as discussed before.

Figure 9.13 shows the gray-scale map of the thermal stress for a coating manufactured with the wet enameling technique.

The figure reports the component of the residual stress in the longitudinal direction, indicated in the figure as x. As for the boundary conditions, a longitudinal length of the material three times that of the original image was considered in order to demonstrate the stress build-up in a wide spread coating.

The stress contours show a steep variation at the interface between the coating and the substrate, as also evidenced in the diagram of Fig. 9.14a where the thermal stress along with a vertical cross-section through the coating, in particular, relative

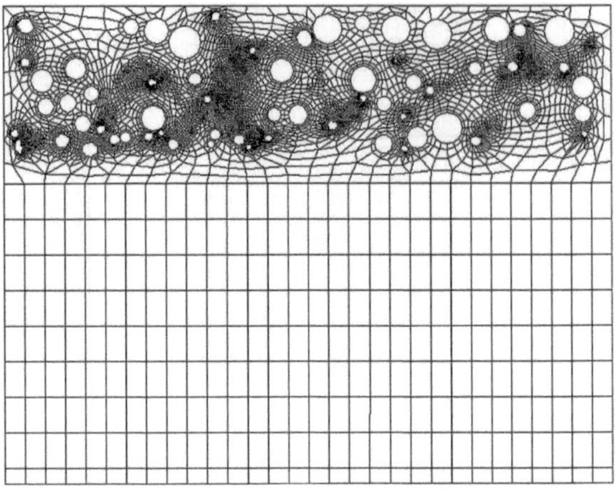

Fig. 9.12 A detail of the F.E.M. mesh

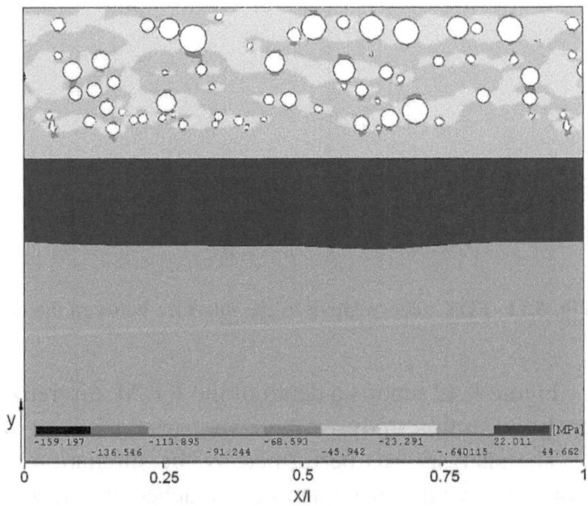

Fig. 9.13 Residual stress for a coating manufactured with wet enameling procedure

to X/l = 0.5, is reported. A zone of tension in the steel substrate and compression stress in the coating are also evidenced.

However, differently from an analogous theoretical coating devoid of bubbles, whose trend of the residual stress is also reported in the figure, higher values of stress exist in the real structure of the coating due to the bubbles. A detail of Fig. 9.13, reported in Fig. 9.14b, shows the zone where the maximum value of stress was measured in the coating.

A similar analysis was conducted on cross-section micrographs relative to the enameled coatings manufactured using the dry enameling technique.

Figure 9.15 reports the gray-scale map of the stress in the longitudinal direction obtained for a representative image.

It can be observed that the maximum values of the residual stress measured in the material are lower than those obtained for the coating produced with the wet

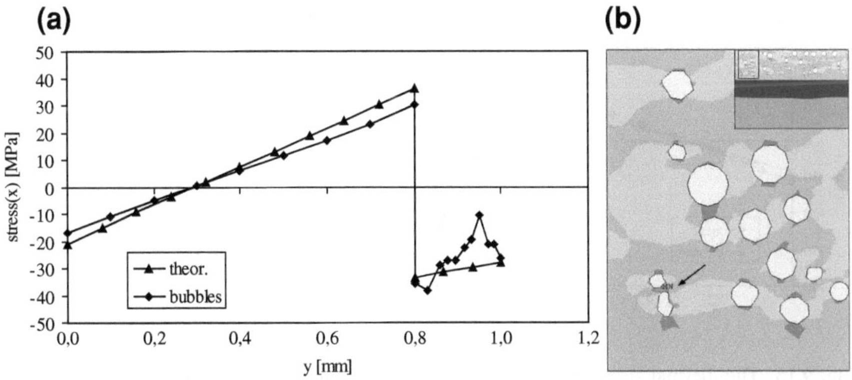

Fig. 9.14 The thermal stress in a wet enameling coating. **a** The stress along with a transversal section **b** a detail of the stress near the bubbles

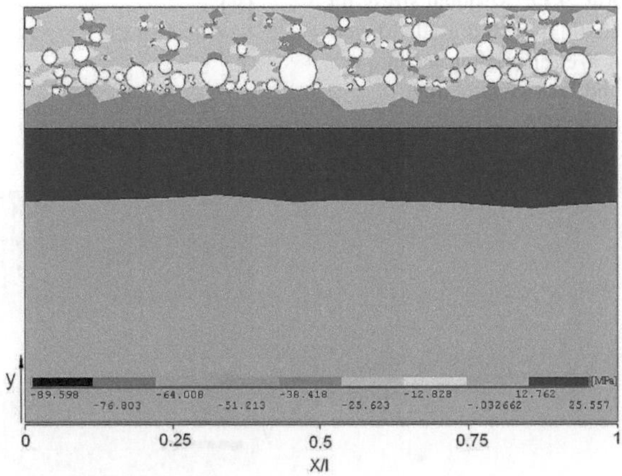

Fig. 9.15 Residual stress for a coating manufactured with the dry enameling technique

enameling manufacture processing. It was expected, since a difference between the thermal expansion coefficient of the two coating systems exists.

Figure 9.16a reports the thermal stress along with a vertical cross-section through the coating, relative to $X/l = 0.5$, while in Fig. 9.16b a detail of Fig. 9.15 is shown, where the zone of the maximum stress measured in the coating is evidenced.

The trend of the stresses is similar to that reported for the previous coating analyzed, and the scattering of the values in the coating is noticeable with respect to the linear trend of a theoretical coating devoid of bubbles.

A further analysis was conducted by taking into account the graded layer between the coating and the substrate. The graded layer was introduced in the models by interposing two slight layers with a thickness of 0.010 mm and linearly variable mechanical and thermal properties between the two constituents. The thickness of the graded layer was derived from measurements on images of higher

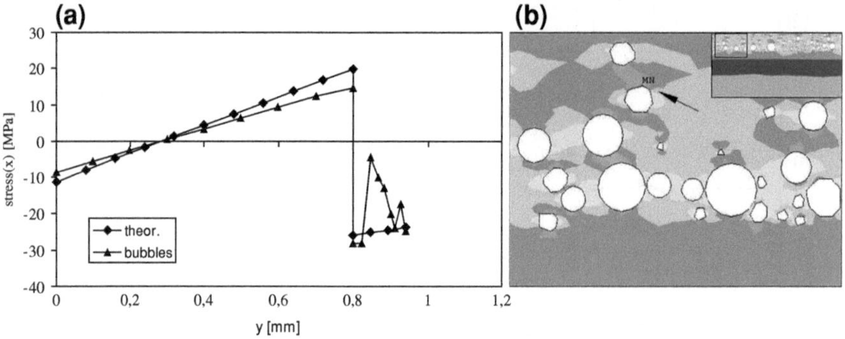

Fig. 9.16 The thermal stress in a dry enameling coating. **a** The stress along with a transversal section **b** a detail of the stress near the bubbles

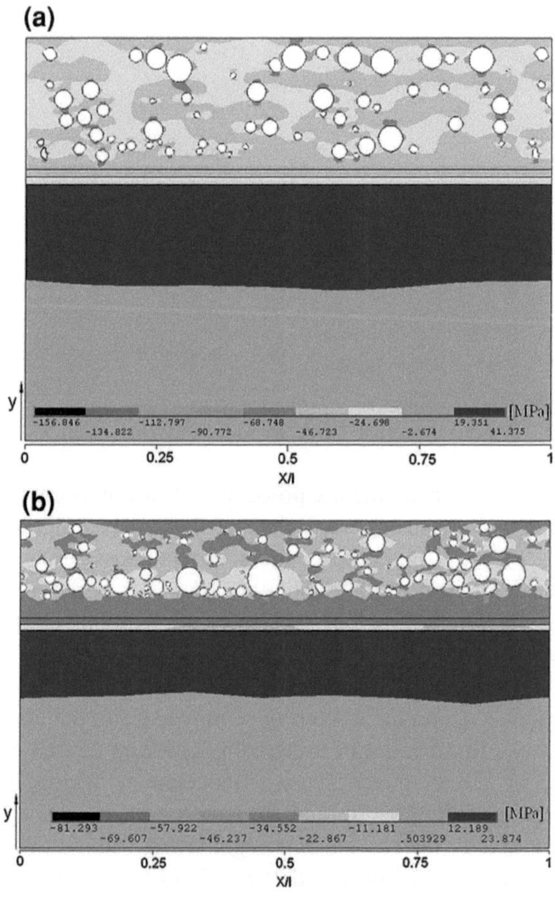

Fig. 9.17 Residual stress for coatings with a graded layer. **a** Wet enameling technique and **b** dry enameling technique

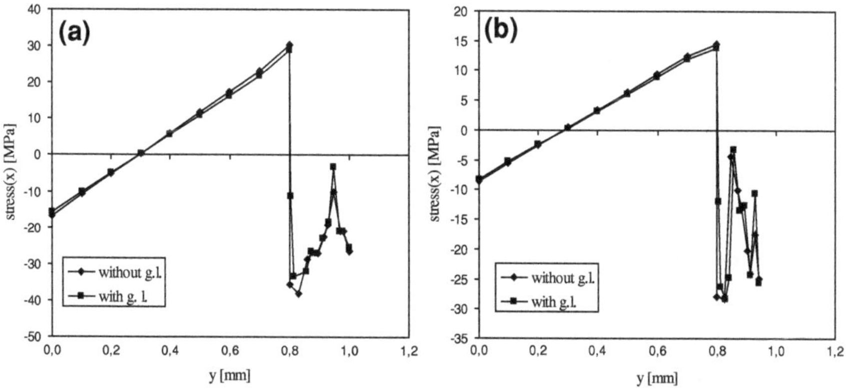

Fig. 9.18 The residual stress along with a transversal section. **a** Wet enameling technique and **b** dry enameling technique

magnification and the values were found similar for the coatings manufactured with the different technologies.

Figure 9.17 shows the gray-scale map of the longitudinal stress obtained in these simulations.

In particular, Fig. 9.17a is relative to the coating manufactured with the wet enameling technique, while Fig. 9.17b is relative to the coating manufactured with the dry enameling technique.

The results show that the presence of the graded layer gives a contribute to the reduction of the maximum values of the thermal stress in the coating, as compared with the previous simulations.

Figure 9.18 shows the trend of the residual stress along with a transversal section, corresponding to $X/l = 0.5$, relative respectively to the coatings manufactured with the two manufacturing technologies.

From the diagrams it can be observed, in particular, that the steep discontinuity in stress at the coating-substrate interface is slightly attenuated when the graded layer is introduced in the models. However, differently from the functionally graded materials where, typically, the graded interlayers are used to mitigate the thermal stresses [17], this occurrence is not particularly evidenced in the coatings analyzed. Among the different factors that contribute to this circumstance, such as geometry and composition, probably, the presence of the bubbles masks the effect on the residual stress of the graded interface between the coating and the substrate.

9.6 Conclusions

In this chapter, a method based on digital images of cross-section micrographs has been applied to the analysis of residual stress in enameled coatings. The results, relative to coatings manufactured with different technologies, highlight the relationship between the local characteristics of the structure and the stresses. In particular, the presence of bubbles in the coating seems to have a significant role on the distribution of the local residual stress while the effect of the graded layer between the coating and the substrate seems to be attenuated and this aspect will be further investigated.

References

1. Andrews, A.I.: Porcelain Enamel. The Garrard Press, Champaign (1961)
2. Vargin, V.V.: Technology of Enamels. Maclaren and Sons, London (1967)
3. Chelli, A., Poletti, R., Pignatti, L., Bruscoli, F., et al.: Composite enameled steel elements for air preheaters and gas-gas heaters: an integrated approach from sheet forming and enamelling to basket assembly. In: XXI International Congress on Porcelain Enamel, Shanghai, pp. 130–158, 18–22 May 2008
4. Chelli, A., Poletti, R., Pignatti, L., et al.: Experimental study of the mechanical and tribological properties of enameled steel plate, (abstract in english, full text in italian). Smalto Porcellanato **3**, 1–27 (2006)
5. Zucchelli, A., Minak, G., Ghelli, D.: Low-velocity impact behavior of vitreous-enameled steel plates. Int. J. Impact Eng. **37**, 673–684 (2010)
6. Zhang, X.C., Xu, B.S., Wang, H.D., et al.: An analytical model for predicting thermal residual stresses in multilayer coating systems. Thin Solid Films **488**, 274–282 (2005)
7. Bengtsson, P., Persson, C.: Modelled and measured residual stresses in plasma sprayed thermal barrier coatings. Surf. Coat. Tech. **92**, 78–86 (1997)
8. Swank, W.D., Gavalya, R.A., Wright, J.K., Wright, R.N.: Residual stress determination from a laser-based curvature measurement. INEEL/CON-99-01176. http://www.osti.gov/bridge/productbiblio.jsp?osti_id=758128(2000) Accessed 11 May 2009
9. Lima, C.R.C., Nin, J., Guilemany, J.M.: Evaluation of residual stresses of thermal barrier coatings with HVOF thermally sprayed bond coats using the Modified Layer Removal Method (MLRM). Surf. Coat. Tech. **200**, 5963–5972 (2006)
10. Wejrzanowski, T., Spychalsky, W.L., Rozniatowsky, K., et al.: Image based analysis of complex microstructures of engineering materials. Int. J. Appl. Math. Comput. Sci. **18**, 33–39 (2008)
11. Von Bradke, M., Gitzhofer, F., Henne, R.: Porosity determination of ceramic materials by digital image analysis–a critical evaluation. Scanning **27**, 132–135 (2005)
12. Ghafouri-Azar, R., Mostaghimi, J., Chandra, S.: Modeling development of residual stresses in thermal spray coatings. Comput. Mater. Sci. **35**, 13–26 (2006)
13. Wang, Z., Kulkarni, A., Deshpande, S., et al.: Effect of pores and interfaces on effective properties of plasma sprayed zirconia coatings. Acta Mater. **51**, 5319–5334 (2003)
14. Russ, J.C.: The image processing handbook. CRC Press, Boca Raton (1995)
15. Abramoff, M.D., Magelhaes, P.J., Ram, S.J.: Image Processing with Image. J. Biophotonics Int. **11**, 36–42 (2004)
16. Hsueh, C.H., Lee, C.S.: Modeling of elastic thermal stresses in two materials joined by a graded layer. Compos. Part B-Eng. **34**, 747–752 (2003)
17. Ravichandran, K.S.: Thermal residual stress in a functionally graded material system. Mater. Sci. Eng. A-Struct. **201**(269), 276 (1995)

Chapter 10
In Situ Investigation and Image-Based Modelling of Aluminium Foam Compression Using Micro X-Ray Computed Tomography

Y. L. Sun, T. Lowe, S. A. McDonald, Q. M. Li and P. J. Withers

Abstract Our understanding of the compressive behaviour of foams can be improved by combining micro X-ray computed tomography (CT) and finite element modelling based on realistic image-based geometries. In this study, the cell structure of an aluminium foam (Alporas^{TM}) specimen and its deformation during continuous low-strain-rate compressive loading are recorded by 'fast' CT imaging. The original 3D meso-structure is used to construct a 3D finite element model (FEM) for simulation. It is observed that local collapse can occur in cells with a wide variety of shapes and sizes, and the compressive strength is determined by the formation and development of the localised deformation bands. The FE prediction of the stress–strain relationship and cell deformation process has reasonable agreement with the experimental observation, especially for the cell-wall collapse corresponding to the plateau in the stress–strain curve. The simulation also indicates that local yielding actually occurs in cell walls well before the plateau regime. The experimental and image-based modelling methods demonstrated here for foams have potential across a very wide range of applications.

List of Abbreviations

CT Computed tomography
FEM Finite element models

Y. L. Sun · Q. M. Li (✉)
School of Mechanical, Aerospace and Civil Engineering, University of Manchester, Sackville Street, Manchester M13 9PL, UK
e-mail: qingming.li@manchester.ac.uk

Y. L. Sun
e-mail: sunyongletl@gmail.com

T. Lowe · S. A. McDonald · P. J. Withers
Henry Moseley X-ray Imaging Facility, School of Materials, University of Manchester, Manchester M13 9PL, UK

F. R. Leta (ed.), *Visual Computing*, Augmented Vision and Reality 4,
DOI: 10.1007/978-3-642-55131-4_10,

10.1 Introduction

Aluminium foams are ultra-light, multi-functional materials, which have been used widely in various engineering applications, such as sandwich construction and energy absorption. In the last 20 years, the compressive behaviour of foams has been studied by means of macro measurement and phenomenological mechanical models based on idealised cell geometries [1, 2]. However, the meso-scale mechanisms determining the compressive strength have not been completely understood due to the complexity of the cell structures and their deformation.

Recent advances in X-ray computed tomography (CT) enable the non-destructive characterisation of the meso-structure and deformation behaviour of a foam using synchrotron [3] and laboratory X-ray sources [4]. The advantage of CT is its capability to capture the internal structure and deformation of a material in three dimensions (3D). However, one limitation of applying this technique during in situ loading is that the test usually has to be interrupted at certain loading stages to allow X-ray scanning. This may introduce stress relaxation and other effects. To overcome this, continuous X-ray CT scanning is used in this study. Moreover, improvements in meshing techniques and the upgrade of computational hardware enable realistic simulations using finite element models (FEM) based on 3D CT images [5]. Such image-based modelling can be applied to isolate the effects of specific physical factors, e.g. cell-wall material properties [6] and strain rate [7], as well as geometrical structural effects. It also can be used to infer the material properties of cell walls using inverse methods [8]. The combination of X-ray in situ CT experiments and image-based modelling can provide detailed information for both qualitative and quantitative analyses.

This work aims at investigating the low-strain-rate compression of a closed-cell aluminium foam by in situ X-ray testing and image-based modelling. First, in situ compression was undertaken during which 3D CT images of the original and successive deformed configurations of the foam were obtained. Then the CT image of the original foam was used as the geometrical basis for a 3D FEM for simulation. The experimental result delineates the deformation mechanism and its effect on the stress–strain relationship of the foam. The numerical prediction provides additional information, e.g. plastic strain distribution in the cell walls.

10.2 Experiment and Modelling

A cylindrical specimen was mechanically cut from a panel of the closed-cell aluminium Alporas™ foam for the in situ compression experiment. The diameter and thickness of the specimen are 20 and 10 mm, respectively. The specimen was scanned using a Nikon Metris CT system housed in a customised bay at the Henry Moseley X-ray Imaging Facility (HMXIF, Manchester, UK), which is capable of fast data acquisition useful for minimising blurring during in situ experiments.

Radiographs were acquired at an accelerating voltage of 75 kV, a current of 220 μA, a voxel size of 15 μm and an exposure time of 250 ms. A 25 kN Deben rig (Deben Ltd, UK) was used to apply the compressive load. Figure 10.1 shows the schematic of the experimental set-up and a resulting X-ray radiograph (projection). Before loading, 2,000 projections were acquired in order to obtain a high quality CT image to enable accurate meshing of the cell structure for numerical modelling. The sample was then compressed at a loading speed of 0.05 mm/min (strain rate of $8.3 \times 10^{-5}\ s^{-1}$) without interruption when scanning was performed. A smaller number of projections, namely 400 projections, were used for each scan during loading to increase the 3D frame rate. Under these conditions, the displacement of the foam surface attached to the loading platen is 0.083 mm per scan, which corresponds to a compressive strain of 0.83 % and sufficiently preserves the imaging quality under the selected resolution, as verified by the visual inspection of the reconstructed images. Thirty scans were conducted over the whole test and reconstructed into 3D volume data using Nikon Metris CT-Pro reconstruction software. Subsequently, the raw CT data were processed by Avizo standard (Visualization Sciences Group, Bordeaux, France) and grey-level segmentation was used to obtain the 3D images of the solid bodies corresponding to different configurations of the foam specimen under the compression.

The CT image of the original cell structure of the foam specimen was used as the basis for the geometry for the 3D finite element model. A virtual transverse slice of the original foam is shown in Fig. 10.2a, from which the pores (black) and the aluminium walls (white) can be distinguished. By segmenting all the slices, the 3D solid body of the foam (see Fig. 10.2b) was obtained and then meshed using the numerical algorithm in ScanIP (Simpleware Ltd, UK). Figure 10.2c and d show the profile of the created mesh comprising about 14.54 million 4-node tetrahedral elements and 0.94 million 8-node hexahedral elements. It should be noted that the voxel size of the CT images was down-sampled to 60 μm and the internal voids smaller than 100 μm were disregarded in the meshing in order to optimise the element size and thus to reduce the element number. As a result, the relative density of the specimen after image processing and meshing is 17.1 %, which is larger than the 13.5 % measured by the grey value method using original CT volume data and the normal values ranging from 8 to 10 % determined by directly weighing a specimen [2]. The discretized structure was then imported into the general purpose FE code Abaqus to create a finite element model. Two rigid platens were placed on the two ends of the modelled foam specimen, which was compressed by the top platen moving downward at a constant speed with the bottom platen being fixed. No constraint was applied directly to the foam. Abaqus/Explicit was used to perform a quasi-static analysis considering the large deformation and the contact condition neglecting friction. An isotropic elastic perfectly plastic material model was adopted for the cell walls. The elastic modulus, Poisson's ratio and yield strength of the cell-wall material were taken to be 68 GPa, 0.33 and 35.5 MPa, respectively [8].

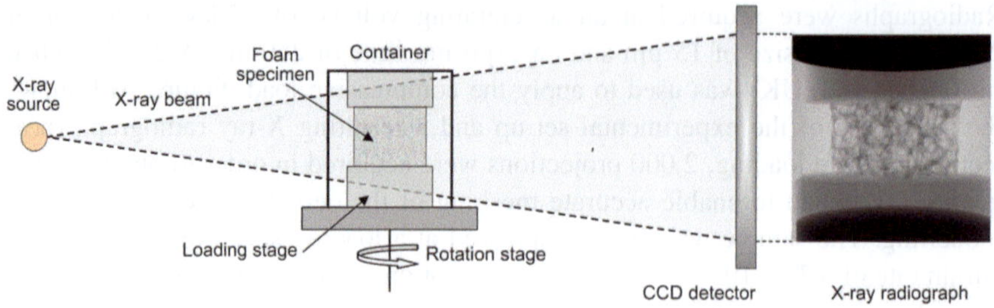

Fig. 10.1 Schematic of the in situ experimental set-up for the foam compression (*left*) and an X-ray radiograph of the scanned foam (*right*)

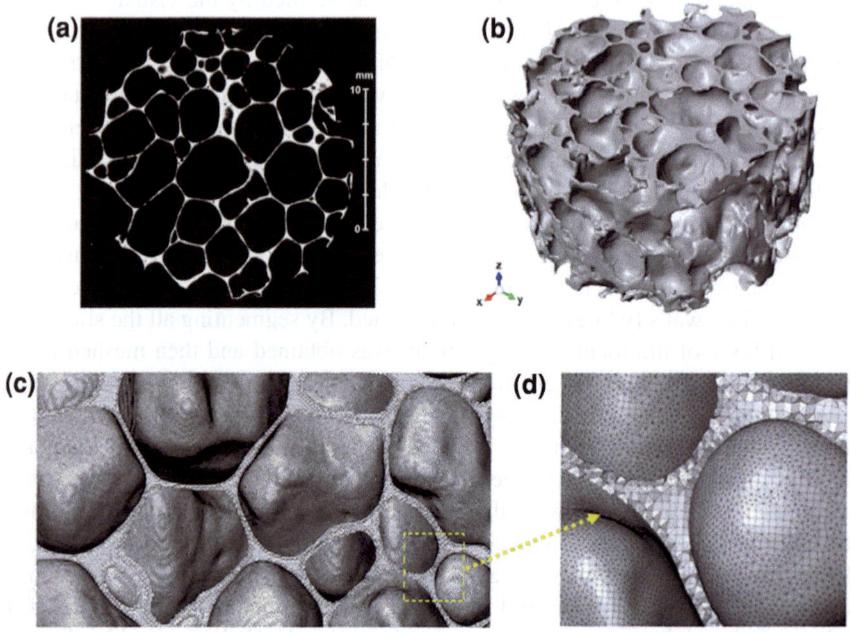

Fig. 10.2 **a** CT slice image in a *transverse plane* of the foam specimen; **b** The 3D image of the foam after the segmentation of the solid portions in the CT volume data; **c** Mesh of the cell structure (view in a *vertical plane*); **d** Enlarged view to show the mesh density

10.3 Results and Discussion

10.3.1 Stress–Strain Relationship

The nominal stress–strain curves obtained from the experiment and simulation are compared in Fig. 10.3. The experimental stress–strain relationship is approximately linear at small strains. However, a nonlinear feature becomes apparent

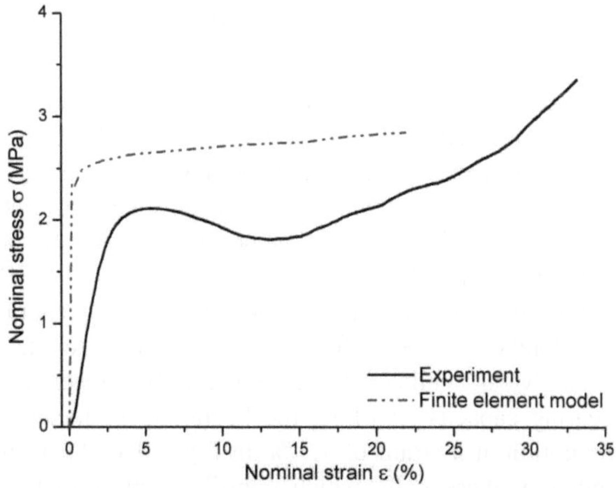

Fig. 10.3 Experimental and numerical stress–strain curves of the compressed foam

when the stress approaches its maximum during the early compression stage, which is followed by a plateau regime showing prolonged deformation at almost constant stress. After considerably more deformation the stress rapidly rises with strain, reflecting the cell interaction in the densification regime. The numerical prediction generally overestimates the stress level and does not precisely mimic the experimental stress–strain curve. Nevertheless, it successfully captures the essential trend of the stress variation, i.e. the stress almost remains constant in the plateau regime. Unfortunately, the simulation of the densification stage cannot be achieved in the present model due to extensive element distortion associated with the large deformations. The predicted modulus of the linear part of the stress–strain curve is also larger than the experimental one, but is close to the unloading modulus of the same foams measured in previous experiments [9]. The discrepancy between the numerical and experimental results can be attributed to the numerical errors caused by the use of linear elements to discretize the cell walls and possibly an insufficient number of elements over the thin solid walls, as well as the inhomogeneous properties of the cell-wall material [10]. To improve the numerical precision further, more elements would be required which was not possible in this case due to the limited computational power available for over 15 million elements. Furthermore, the increased relative density after meshing and the lack of reliable homogenised material models for the cell walls also hinder more accurate prediction. Nevertheless, the effect of the real meso-scale geometry on the compressive behaviour has been captured in the FE model. Therefore, the current simulation should be regarded as qualitative and capable of providing insights into the compressive behaviour of foams.

10.3.2 Deformation Mechanisms

The foam has an extremely complex cell structure with geometrical imperfections and hierarchical features. From the virtual CT slices shown in Fig. 10.4a and b, it is clearly seen that the cell shape deviates from regular ellipsoids and polyhedra. Moreover, the cell walls themselves are somewhat porous, being significantly thicker at the interconnecting portions. Figure 10.4a and b show the cell deformations within the central cross-section of the 3D volume. The compressive deformation is markedly inhomogeneous, being characterised by the formation and development of localised deformation bands. As the compressive strain increases, some 'weak' cells collapse prematurely, presumably due to their inferior load bearing capacity, marking the beginning of localised deformation (see the configuration at a strain of 7.7 % in Fig. 10.4). Deformation then becomes concentrated in these cells which contribute most to the subsequent increase in the compressive strain, indicating the development of a deformation band (see the configuration at a strain of 17.8 % in Fig. 10.4). This process continues until new deformation bands are formed. Such deformation on the cell level corresponds to the plateau regime in the stress–strain curve observed in the macro measurement. Therefore, in common with previous observations [11], the meso-scale cause of the plateau stress, which represents the compressive strength of the foam, is the progressive cell collapse occurring in the localised deformation bands.

The prematurely collapsing 'weak' cells play a crucial role in the localised deformation mechanism. However, the exact cause of the premature collapse has not been unequivocally identified. Based on the observation of cell deformation within the cross-section, Bart-Smith et al. [11] identified two critical cell morphologies, i.e. ellipsoidal cells with T-shaped wall intersections and cells with appreciably curved walls, for the weak cells, but these morphologies cannot be easily distinguished in the present experimental observations. Our results reveal that the weak cells susceptible to collapse actually have a wide variety of shapes and sizes. Consequently, existing simple (mostly 2D) geometric criteria are unlikely to describe them, especially when their 3D nature is considered. It is questionable whether a 2D geometry can describe the 3D cell structure of a foam, since the 2D geometric characteristics of one foam cell may change significantly from one cross-section to another. For instance, a 'small cell' observed in one cross-section may be part of a large cell and the 2D curvature does not necessarily represent the 3D one. Therefore, the observation of 3D cell deformation is important in establishing a better understanding of the local collapse behaviour.

The deformation bands in the central XZ and YZ planes are shown in Fig. 10.4a and b as the dotted lines. The comparison between them indicates that the location and orientation of the bands depend on the plane selected. The band in the XZ plane is close to the top and is slightly inclined. By contrast, the band in the YZ plane is closer to the bottom and is significantly inclined. The cells outside the bands essentially retain their original shape, even when the compressive strain has reached 17.8 %. To inspect the 3D features of these deformation bands, translucent

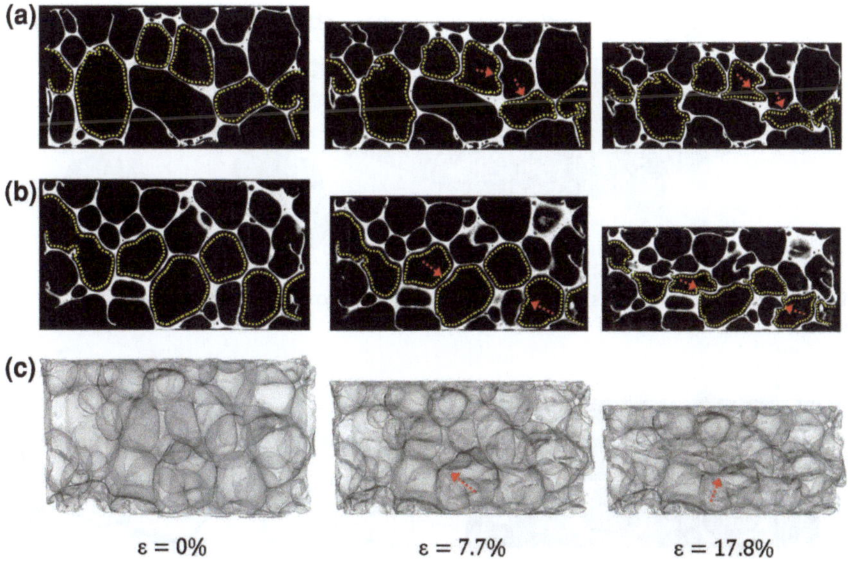

Fig. 10.4 Configurations of the foam specimen at different nominal strains: **a** Cross-section in the central XZ plane; **b** Cross-section in the central YZ plane; **c** Translucent foam body to show the 3D deformation of internal cells. The *dotted lines* indicate the cells in the localised deformation bands and the *arrows* indicate the collapse modes of cell walls

images of the foam body are presented and compared, as shown in Fig. 10.4c. It is evident that shear deformation across cells is dominant on the cell level, see the cells indicated by the arrows in Fig. 10.4c, whereas both bending and buckling occur at the smaller scale associated with cell walls, see the portions indicated by the arrows in Fig. 10.4a and b.

It reveals that the load distribution across the cells is very complicated even when the foam is subjected to uniaxial compression. The 3D deformation of two centrally located cells is shown in Fig. 10.5 and complex morphological changes are clearly seen. These cross-sectional and 3D observations confirm that premature cell collapse depends not only on the cell morphology but also on the actual load the cell experiences.

Figure 10.6 shows a comparison between the experimental observation and the numerical prediction. It is seen that the FE model predicts well the cell deformation, including the location and mode of the collapse of cell walls. Some fine features may be lost in the simulation because of the limited precision of the reconstruction and the meshing. Nevertheless, the simulation captures the essential deformation mechanisms and allows further computational analysis. For instance, the equivalent plastic strain distribution in the cell walls can be obtained from the simulation, and it indicates that extensive plastic deformation has already occurred in some locations at a strain of 0.5 %, as shown in Fig. 10.7, which is well before the plateau stage is reached. Such local yielding at small strain is probably the

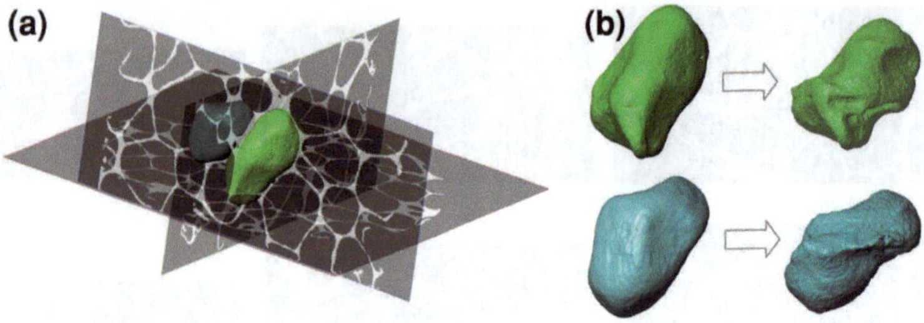

Fig. 10.5 **a** Two typical cells located in the centre of the foam specimen; **b** 3D deformation of the central cells at a strain of 12.0 %

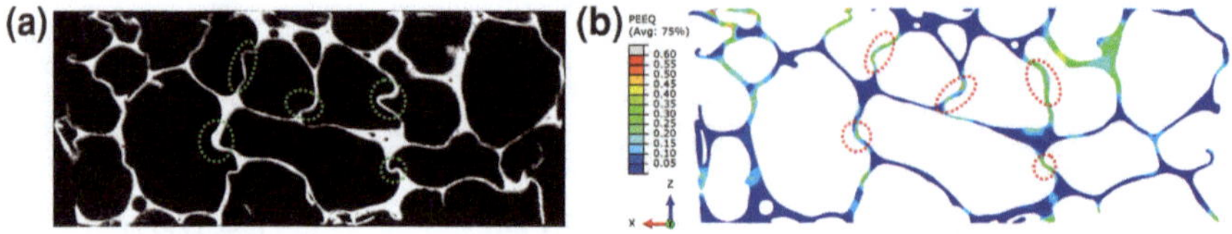

Fig. 10.6 Comparison of the cell deformation in the central XZ plane at a strain of 17.8 %: **a** Slice image from CT data; **b** Numerical result (contour of the equivalent plastic strain in the deformed configuration). The extensively deformed local portions of the cell walls are indicated by the dotted ellipses

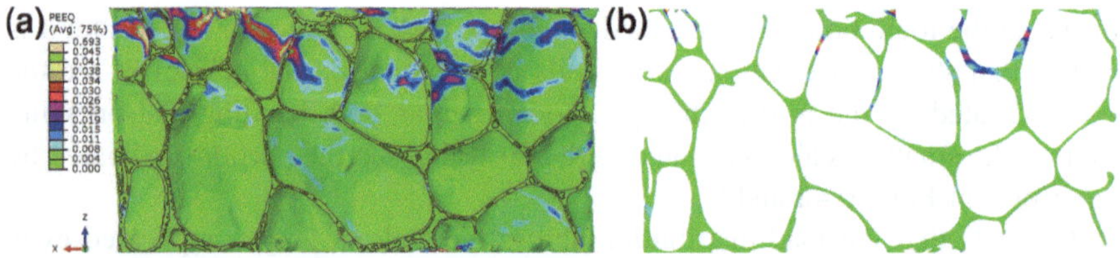

Fig. 10.7 Plastic strain distribution in the cell walls at a strain of 0.5 %: **a** Half of the solid body viewed in the Y direction; **b** Cross-section in the central XZ plane

cause of the discrepancy between the tangent modulus and unloading modulus of the linear part of the stress–strain curves observed in previous experiments [9].

10.4 Conclusions

In situ X-ray compression and image-based modelling are used to investigate the low-strain-rate compressive behaviour of a closed-cell aluminium foam. According to the experimental observation, the compressive strength is mainly determined by cell collapse in the localised deformation bands. The CT images reveal that some cells exhibit premature collapse due to morphological effect and complex load distribution, and both bending and buckling occur in the walls of the 'weak' cells having various shapes and sizes. The finite element simulation based on the real cell structure extracted from CT images predicts the stress–strain curve and cell deformation in a reasonable agreement. The numerical prediction can be improved by developing a more reliable material model for the cell walls and reducing numerical errors associated with the meshing. The simulation indicates that significant local yielding occurs well before the plateau regime is reached.

Acknowledgments The authors would like to acknowledge the assistance given by the IT Services and the use of the Computational Shared Facility at The University of Manchester. The supports from the Engineering and Physical Science Research Council (EPSRC) grants EP/F007906/1 and EP/F028431/1 and Royal Society grant JP100958 are also acknowledged. The first author is grateful for the PhD scholarship from the School of Mechanical, Aerospace and Civil Engineering, The University of Manchester.

References

1. Gibson L.J., Ashby, M.F.: Cellular Solids: Structure and Properties. Cambridge University Press, Cambridge (1997)
2. Ashby, M.F., Evans, A.G., et al.:(2000) Metal Foams: a Design Guide. Elsevier, Netherlands
3. Maire, E., Babout, L., et al.: Recent results on 3D characterisation of microstructure and damage of metal matrix composites and a metallic foam using X-ray tomography. Mater. Sci. Eng. A **319–321**, 216–219 (2001)
4. McDonald, S.A., Mummery, P.M., et al.: Characterization of the three-dimensional structure of a metallic foam during compressive deformation. J. Microsc. **223**(2), 150–158 (2006)
5. Young, P.G., Beresford-West, T.B.H., et al.: An efficient approach to converting three-dimensional image data into highly accurate computational models. Philos. Trans. R. Soc. A: Math. Phys. Eng. Sci. **366**(1878), 3155–3173 (2008)
6. Jeon, I., Asahina, T., et al.: Finite element simulation of the plastic collapse of closed-cell aluminum foams with X-ray computed tomography. Mech. Mater. **42**(3), 227–236 (2010)
7. Vesenjak, M., Veyhl, C., et al.: Analysis of anisotropy and strain rate sensitivity of open-cell metal foam. Mater. Sci. Eng. A **541**, 105–109 (2012)
8. Jeon, I., Katou, K., et al.: Cell wall mechanical properties of closed-cell Al foam. Mech. Mater. **41**(1), 60–73 (2009)
9. Andrews, E., Sanders, W., et al.: Compressive and tensile behaviour of aluminum foams. Mater. Sci. Eng. A **270**(2), 113–124 (1999)
10. Simone, A.E., Gibson, L.J.: Aluminum foams produced by liquid-state processes. Acta Mater. **46**(9), 3109–3123 (1998)
11. Bart-Smith, H., Bastawros, A.F., et al.: Compressive deformation and yielding mechanisms in cellular Al alloys determined using X-ray tomography and surface strain mapping. Acta Mater. **46**(10), 3583–3592 (1998)

Part IV
Related Problems

Chapter 11
Multiple Plane Detection Method from Range Data of Digital Imaging System for Moving Robot Applications

Jeong-Hyun Kim, Zhu Teng, Dong-Joong Kang and Jong-Eun Ha

Abstract Imaging system using CCD sensors for automatic navigation of intelligent robot is a central element to recognize and interact with the surrounding environment. Specifically, finding a planar surface on 3D space is very important for efficient and safe operation of a mobile robot. In this chapter, a noise rejection filter is introduced for defining planar surfaces to reduce the voting of noisy data. We test the normal directions of two arbitrary planes in a small region, which are determined by three vertexes of a triangle and its rotation. If the angle of two normal directions is lower than a given threshold, it is voted into the Hough parameter space. This method is similar to a noise rejection filter to verify the planarity of local planes. We can get accurate parameters of the plane in RHT because most noises and nonplanar data cannot vote into the Hough parameter space. We use a scan window to vote locally. The scan window explores all regions by changing the window size. The window operation improves the accuracy of plane detection because the plane is locally consistent and increases

J.-H. Kim
Realhub research institute, Realhub corporation limited, Jangjeon-dong, Geumjeong-gu, Busan 609-735, Korea
e-mail: feelmare@daum.net

Z. Teng
School of Computer and Information Technology, Beijing Jiaotong University, No.3 Shang Yuan Cun, Hai dian District, 100044, Beijing, China
e-mail: zteng@bjtu.edu.cn

D.-J. Kang (✉)
School of Mechanical Engineering, Pusan National University, Jangjeon-dong, Geumjeong-gu, Busan 609-735, Korea
e-mail: djkang@pusan.ac.kr

J.-E. Ha
Deptartment of Automotive Engineering, Seoul National University of Science and Technology, Gongnung-dong 232, Seoul, Korea
e-mail: jeha@seoultech.ac.kr

F. R. Leta (ed.), *Visual Computing*, Augmented Vision and Reality 4,
DOI: 10.1007/978-3-642-55131-4_11,

the search speed for finding planes. Finally, the performance of the algorithm for real range data obtained from a stereo imaging system has been verified.

Keywords Digital imaging system · 3D range data · Plane detection · Randomized Hough transform · Mobile robot

List of Abbreviations

CHT	Combinatorial Hough transform
DGHT	Dynamic generalized Hough transform
HT	Hough transform
IRHT	Iteractive randomized Hough transform
KIAT	Korea Institute for Advancement of Technology
LUT	Look up table
MEST	Ministry of Education, Science Technology
NRF	National Research Foundation of Korea
PDC	Plane detection cell
RHT	Randomized Hough transform

11.1 Introduction

Finding a planar region in 3D space is important for safely driving and operating a mobile robot or a moving system that can freely drive. A robot operating in an unknown 3D space must identify its surroundings before the system can conduct its assigned tasks. Biped or wheel-based robots should be able to recognize obstacles within their area of motion and avoid or drive over detected obstacles where possible. In the presence of a staircase, a robot should be able to recognize the step planes and traverse them. In order to operate without falling in an unknown 3D space, it must be able to use the available information about 3D depth to recognize planar regions without obstacles in the direction of movement.

Planar space detection has many fields of application. A robot moving around indoors should be able to recognize the planar floor region on which it will drive. Also, the inclination of this plane with respect to the direction of gravity should not be too great, to prevent the robot falling while it is moving. Planar space recognition can be used in recognizing obstacles, such as walls, a table on which objects are located, or the plane of a table. It can also be used to find the centerline of road on which a robot can drive. Recognizing the step planes is an important application for the movement of humanoid biped robots [1].

Recently, a method was proposed which detects regions in real-time for driving mobile robots. Based on the Hough Transform (HT), the method was applied to biped robots such as ASIMO of Honda, which walked on steps and avoided obstacles successfully. Using the HT, the *X-Y-Z* data in 3D space is transformed into another parameter space, with ρ-θ-φ, representing the vertical distance from the origin and the rotation angle on the plane for voting. The corresponding planes are detected by the peaks in the voting space.

Although the method produces segmentation results for range data with some noises and complex planes for an indoor environment, it requires many impractical procedures such as excessive memory for voting, decisions about the optimal voting size, difficulties in fixing the peak locations in the voting space, and many additional processes which need a high computer processing speed [2]. In order to solve these problems, many improved methods have been proposed such as Combinatorial HT (CHT) [3], Randomized HT (RHT) [4, 5], Probabilistic HT (PHT) [6], and Dynamic Generalized HT (DGHT) [7].

RHT is proposed to solve the computing time problem of HT. It doesn't vote all points in an image into Hough space but votes on the geometrical parameters which are calculated by selecting several points randomly. However, sampling range data that exists on different planes or includes noisy samples on a nonplane surface result in many local maxima in the parameter voting distribution of the Hough space. Also, several distributions are mixed and thus the peaks can be very similar. In a mixed and overlapped distribution of peaks in Hough parameter space, it is difficult to segment a plane reliably.

Kang et al. proposed the plane detection cell (PDC) for reducing the effect of noises and outliers [8]. PDC is a cell of circle type used to test the planarity. The normal vector of a triangle is inscribed in a small circular region, the triangle is rotationally sampled and a series of inscribed triangles having different normal vectors is generated. The direction vectors of these generated triangles are used to test the planarity of the small circular region.

This method is effective in a local region detection but it is not adequate for global plane detection and segmentation and calculating the planarity at all the grid point is too time-consuming.

We test only the normal directions of two planes, which are determined by three vertices of a triangle on range data and arbitrary rotation of it. If the angle of two normal detections is lower than a given threshold it is voted into Hough parameter space. It takes less time than PDC and also detects the global planes, because it can provide the global voting through the local evaluation of data. First, we start with a scan window to vote locally, then, the scan window explores all regions as its size grows. This method improves the detection performance because the planes are locally consistent. We obtained 3D range data from a stereo imaging system and experimented with it in various environments.

The remainder of this chapter is organized as follows: Sect. 11.2 describes the RHT method for plane detection. Section 11.3 presents details of the proposed method. Section 11.3.1 describes a method using random sampling from accuracy evaluation of local planar region, Sect. 11.3.2 shows a search method using a scan window, and Sect. 11.3.3 provides a Look-Up-Table for fast processing. Section 11.3.4 explains the application of Iterative RHT to reduce the effect of noises and multiple planes. The experimental results are shown in Sects. 11.4 and 11.5 gives the conclusions.

11.2 Randomized Hough Transform

The Hough transform method has the advantage that it is robust to noise when applied to detect a geometric shape in inputted 2D/3D data. However, the computational cost and required memory size are very large. To cope with this problem, a method called randomized Hough transform has been introduced [6, 9].

In the original Hough transform method, each point on the input image is transformed into a Hough parameter space, which is why there are so many points for voting. However, Randomized HT votes for geometrical parameters which are calculated by selecting several points randomly. Figure 11.1 shows the difference between RHT and HT in case of detection of 3D spatial planes.

$$Ax + By + Cz + D = 0 \tag{11.1}$$

RHT selects three points randomly for detecting a plane, and the plane parameters (A, B, C, D), which represent a plane calculated from the three points, is voted into the Hough parameter space. If the three points denoted by ($x1$, $y1$, $z1$) ($x2$, $y2$, $z2$), and ($x3$, $y3$, $z3$) are set on a plane, parameters A, B, C, D can be represented by Eq. (11.2).

$$\begin{aligned} A &= \begin{vmatrix} 1 & y1 & z1 \\ 1 & y2 & z2 \\ 1 & y3 & z3 \end{vmatrix} \quad B = \begin{vmatrix} x1 & 1 & z1 \\ x2 & 1 & z2 \\ x3 & 1 & z3 \end{vmatrix} \\ C &= \begin{vmatrix} x1 & y1 & 1 \\ x2 & y2 & 1 \\ x3 & y3 & 1 \end{vmatrix} \quad D = -\begin{vmatrix} x1 & y1 & z1 \\ x2 & y2 & z2 \\ x3 & y3 & z3 \end{vmatrix} \end{aligned} \tag{11.2}$$

The parameters A, B, C, D are difficult to define in a limited range because they can have fractional values. We need to change this to a spherical coordinate system. Figure 11.2 shows the spherical coordinate system, where ρ is the distance between a plane and the origin, φ is the angle with respect to the x-axis, and θ is the angle with respect to the z-axis. We can define the range of parameters in the changed coordinate system:

$$\begin{aligned} &\rho \geq 0 \\ &0 \leq \theta \leq \pi \\ &0 \leq \varphi \leq 2\pi \end{aligned} \tag{11.3}$$

where ρ, φ, θcan be obtained by Eq. (11.4), where x, y, z is a point in the Cartesian coordinate system and they are denoted by $x = A/D$, $y = B/D$, $z = C/D$.

$$\rho = \sqrt{x^2 + y^2 + z^2}, \quad \theta = \cos^{-1}\left(\frac{z}{\rho}\right), \quad \varphi = \tan^{-1}\left(\frac{y}{x}\right) \tag{11.4}$$

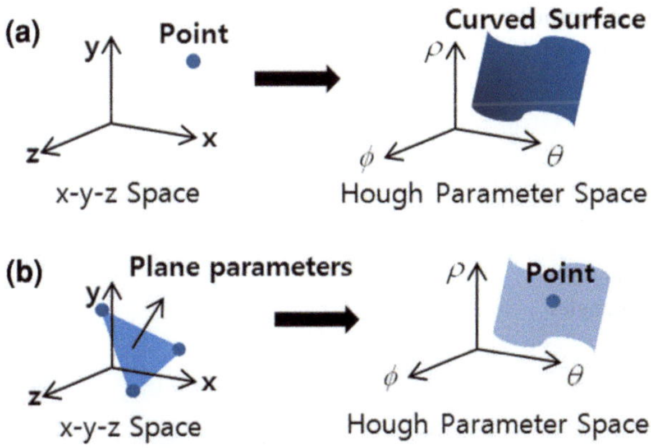

Fig. 11.1 Comparison of HT and RHT. **a** For Hough transform, a point in *xyz* space is transformed to a curved surface in Hough parameter space; **b** For randomized HT, plane parameters in *xyz* space are transformed to a point in Hough parameter space. Three points in 3D space can represent a plane via Eq. (11.1). The 3D point (x, y, z) is on the plane, the parameters A, B, C represent the normal vector of the plane and D is the scale of the normal vector

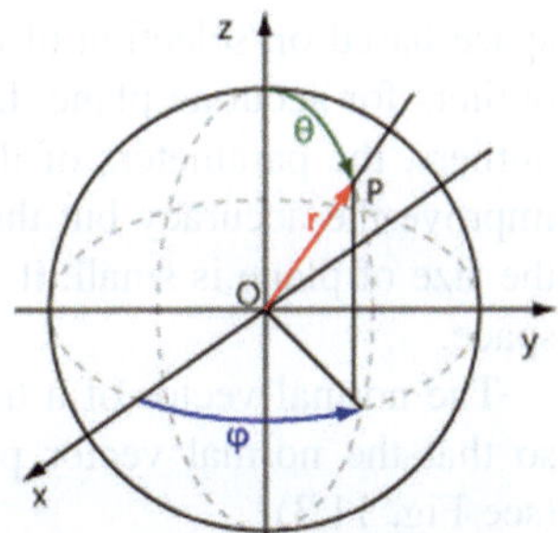

Fig. 11.2 Spherical coordinate system

Conversely, the Cartesian coordinates may be retrieved from the spherical coordinates by Eq. (11.5).

$$x = \rho \sin\theta \cos\varphi, \quad y = \rho \sin\theta \sin\varphi, \quad z = \rho \cos\theta \tag{11.5}$$

11.3 Multiple Plane Detection

11.3.1 Planarity Evaluation from Sampling of Triangular Point Sets

The 3D range data is the coordinate information, measured from the camera origin, of various planar and nonplane surfaces in 3D, including many noises and outliers. The random sampling method of the conventional RHT votes for Hough parameter

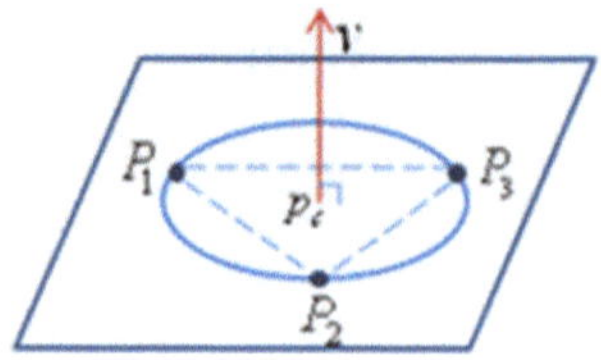

Fig. 11.3 The plane normal vector by three vertices of a triangle

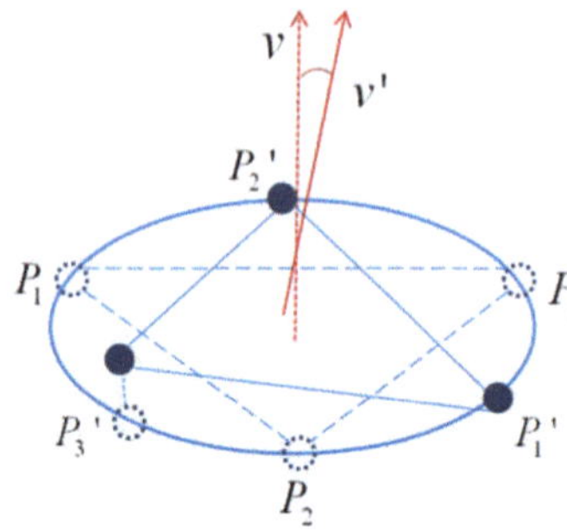

Fig. 11.4 Evaluating the planarity by two triangles

space based on selection of excessive samples in order to reduce the influence of outliers for accurate plane detection. Compared to the portion of the surface with outliers, the parameters of the plane are much more consistent. Many samplings improve the accuracy but the processing time is also much increased. And when the size of plane is small, it is difficult to find the plane peak in Hough parameter space.

The normal vector of a triangle in PDC is inscribed in a small circular region so that the normal vector passes through the circumcenter area of the triangle (see Fig. 11.3).

The triangle is rotationally sampled with respect to the center position of the circular region, and a series of inscribed triangles having different normal vectors are generated. The direction vectors of these generated triangles are normalized and the median direction of the normal vector is then used to test the planarity. If the planarity of the circular region is lower than a given threshold, the point won't be voted into the Hough parameter space. Therefore, this method can filter noises and outliers instead of voting them into the Hough parameter space and it is effective for detection of local regions. But it cannot detect global plane because the method calculates the planarity at a single point. So we use the PDC as a filtering method to reduce the excessive sampling of RHT.

We compare the normal direction of a few planes which are determined by three vertices of a triangle and the others rotated with respect to this triangle (see Fig. 11.4). If the angle between two normal directions is lower than a given threshold it is voted into Hough parameter space. The local evaluation for the planarity acts to filter noises and outliers that are not voted into Hough parameter space. Therefore, there are only a few data samplings, resulting in less computing time than PDC.

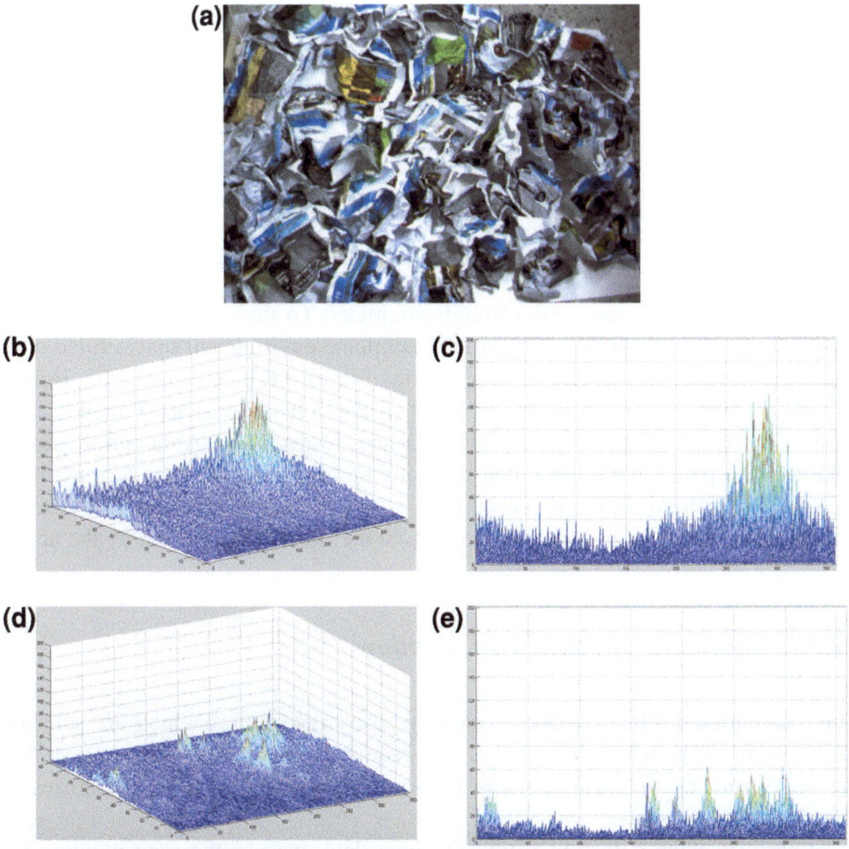

Fig. 11.5 The Hough parameter space of the conventional RHT and the proposed RHT method with local plane filter for a nonplane surface

The vector v normal to the triangular plane obtained by the three spatial points is calculated by Eq. (11.6).

$$v = (P_2 - P_1) \times (P_3 - P_1) \tag{11.6}$$

The vector v is normalized by the expression in Eq. (11.7), and the angle between the two vectors can be obtained by using an inner product Eq. (11.8), if we use two triangular point sets.

$$n = \frac{v}{\|v\|} \tag{11.7}$$

$$\alpha = n \cdot n' \tag{11.8}$$

Figure 11.5 shows the difference between the conventional RHT method and the proposed RHT method using local evaluation filter of triangular sampling

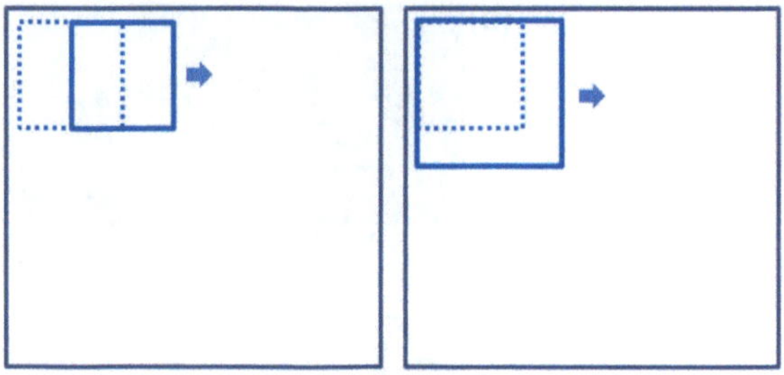

Fig. 11.6 The window scans all regions while changing the window size

method in the voting space of the Hough parameter. To show the filtering effect for nonplanar and noisy surfaces, construct a nonplane environment that is filled with crumpled chapter as shown in Fig. 11.5a. Figure 11.5b–e shows the Hough parameter space, where (b, c) is the result of the conventional RHT method and (d, e) is the result of the proposed RHT method, respectively. Although the range data have no plane, RHT has voted on many parameters and count of maximum peak is 158. In contrast, the proposed RHT's voting count is much smaller and the maximum peak is 61. Obviously, the proposed method is less affected by the noise.

11.3.2 Speed Up via the Scan Window

The conventional RHT involves voting on many data points which are selected randomly from the whole image region. It requires a lot of sampling to prove that the peak from the interesting shape is more salient than those of the noises in Hough parameter space.

In case of plane detection from 3D range data, we use the scan window for accurate and rapid detection of the surface by using the local consistency for data of planar surface. Plane data points have the locality property, because the data points in local region have higher probability to be a plane than far distance data. The scan window is a similar method to that of searching face in AdaBoost [10].

Window that shifts fixed number of pixels scans the whole area. The size of the scan window is initially small and is increased to the image size (see Fig. 11.6). A certain number of data points in each window region are sampled and voted on.

The scan window of various sizes can reflect the locality and the overlapping region of each window provides a more significant peak of the plane parameter than the conventional searching method.

11.3.3 Look Up Table

We compare the normal vector from the initial three vertices with the vector from the rotated three vertices, for planarity evaluation. Equation (11.9) is needed for rotation of the three vertices.

Table 11.1 Some angles for data sampling

1st	2nd	3rd	4th	5th	6th	7th	8th	9th	10th
180°	205°	155°	230°	130°	255°	105°	280°	80°	305°

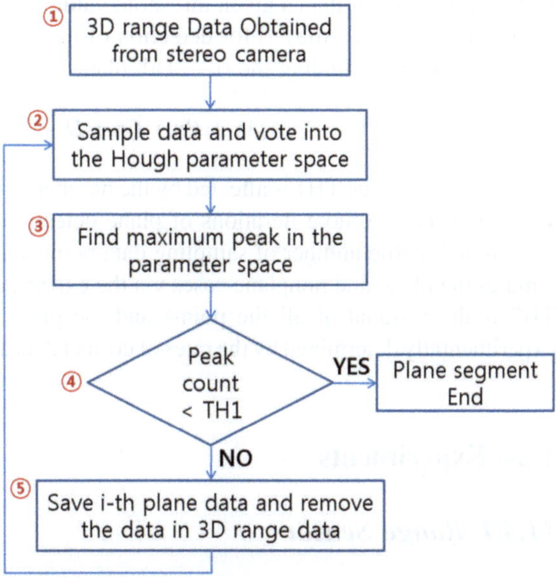

Fig. 11.7 The method of plane segment using IRHT

Because many samplings are required for accurate plane detection, the rotation of three vertices must be repeatedly calculated. Also, we may fail to get the 3D coordinates from the measurement error of the sensor system after a rotation for the original three points. In this case, we try to get the 3D coordinates from another angle. Table 11.1 shows sequential attempts of a few rotation angles.

Therefore, we can previously save the calculated result of trigonometric function into the LUT (Look up table) to avoid repeated calculation of the functions. Although additional memory is necessary for building the LUT, it is indispensable for real-time algorithmic processing.

11.3.4 Plane Segmentation

IRHT (Iterative Randomized Hough Transform) [11] is introduced for plane segmentation. Figure 11.7 presents the plane segmentation steps. IRHT is a segmentation method using sequential elimination of maximum peak data from whole image to reduce the influence of noises and other candidate shapes.

In step ①, the 3D range data is obtained from the stereo camera. The ρ, θ, and φ values of the sampled data are voted into Hough parameter space, as in step ②.

In step ③, we find the maximum peak in Hough parameter space. The peak represents the normal direction of the biggest plane in the 3D range data. In step ④, if the voting count of the peak is lower than TH1, we determine that there is no plane and the plane segmentation algorithm is ended. In ⑤, we find the plane data in the 3D range data by using Eq. (11.9), which is derived from the plane equation. If the value is lower than a given threshold value TH2, the data is saved into the ith plane data and it is removed for detection of the $(i + 1)$th plane. Again, in ②, the algorithm involves finding the $(i + 1)$th plane from the remaining 3D range data.

$$|Ax + By + Cz + D| < \mathrm{TH2} \tag{11.9}$$

A threshold value TH1 is affected by the number of planes detected. The lower the value of TH1, the more iterations of plane detection there are. Therefore, TH1 is determined by the number of sampling data points and the voting value that differentiates the plane and nonplane cases via the experiments. And, the threshold value TH2 is the residual of all the points and the plane we have detected. It is also experimentally determined by the rates of correct data and noises via the experiments.

11.4 Experiments

11.4.1 Range Sensor

We have implemented the proposed method on a Core™2 CPU 2.0 GHz computer and used a Bumblebee stereo imaging system [12] for obtaining the 3D range data.

The quality of the range data from the bumblebee stereo camera is lower than that from the active sensor such as laser scanner and it cannot obtain the range data in a nontexture region, because it fails to find the correlations between the left and right images. However, the range data from the stereo camera can be obtained from a low-cost computer practically in real-time.

Figure 11.8a, b shows the left and right images of the bumblebee stereo camera and Fig. 11.8c shows the depth information from the horizontal disparity between the two images. The region close to the camera is dark, and the bright region is distant from the camera.

In the right-top region of Fig. 11.8c, the gray color region doesn't have any disparity information because there is no texture.

11.4.2 Comparison Results

In the conventional RHT method the sampling times are 2,000,000 per iteration and planes are detected with the TH1 value of 30 for the IRHT algorithm in Fig. 11.7.

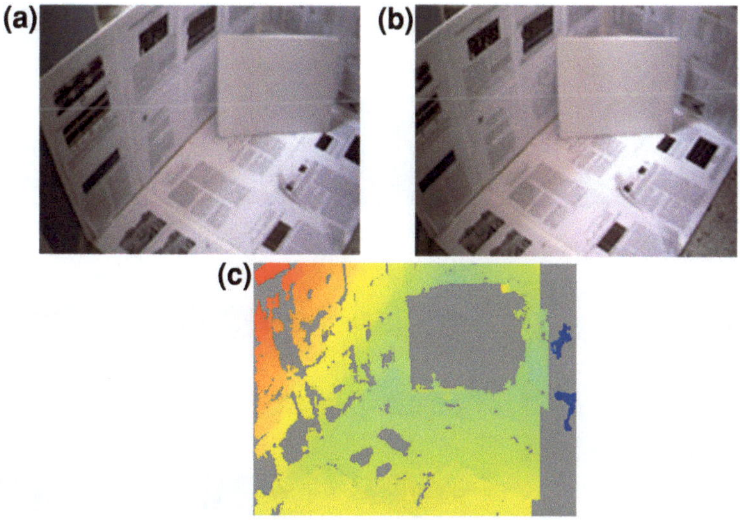

Fig. 11.8 The *left* and *right* image of the stereo camera and 3D range image **a** Left image; **b** Right image; **c** 3D range image

In the proposed method, the initial size of the window is 16 × 12, the shift size is 4 pixels, and the window size is increased by 10 pixels. The sampling times are 1,637,305 and the TH1 value is 100. The parameter is voted on only if the angle of two normal vectors is less than 10°. Also, the weight of the candidate normal vector is set to a higher value when it has better planarity. The following rule applies: voting count = current voting count + (angle threshold – angle of two normal vectors). For example, if the angle is 0°, which is the best planarity, it is voted as having the highest value (10) and if the angle is 9°, which is the worst planarity, it is voted as having the lowest value (1). The parameter value is voted as the mean of the two normal vectors. The TH2 value for removing the plane data is 0.4 in both the conventional RHT method and the proposed method. We can get these threshold values according to our experiments.

Figures 11.9–11.10 and Tables 11.2–11.3 show the comparison results of the conventional RHT method and the proposed method. The top images are obtained from the bumblebee stereo camera. These are the left, right, and 3D range images, respectively. The middle ones are the results of plane detection, which are represented by the segmentation image and Hough parameter space. The colors of the middle left images denote the result of the detected planes and the right graphs show the result of the voting count of the Hough parameter space during each iteration of IRHT. The graph of the Hough parameter space only represents the θ and φ values and not the ρ value. The bottom table represents the ρ, θ, φ values for plane detection.

Figure 11.9 and Table 11.2 show an experiment for one plane detection on an uneven flat surface. The proposed method detected the two planes instead of one

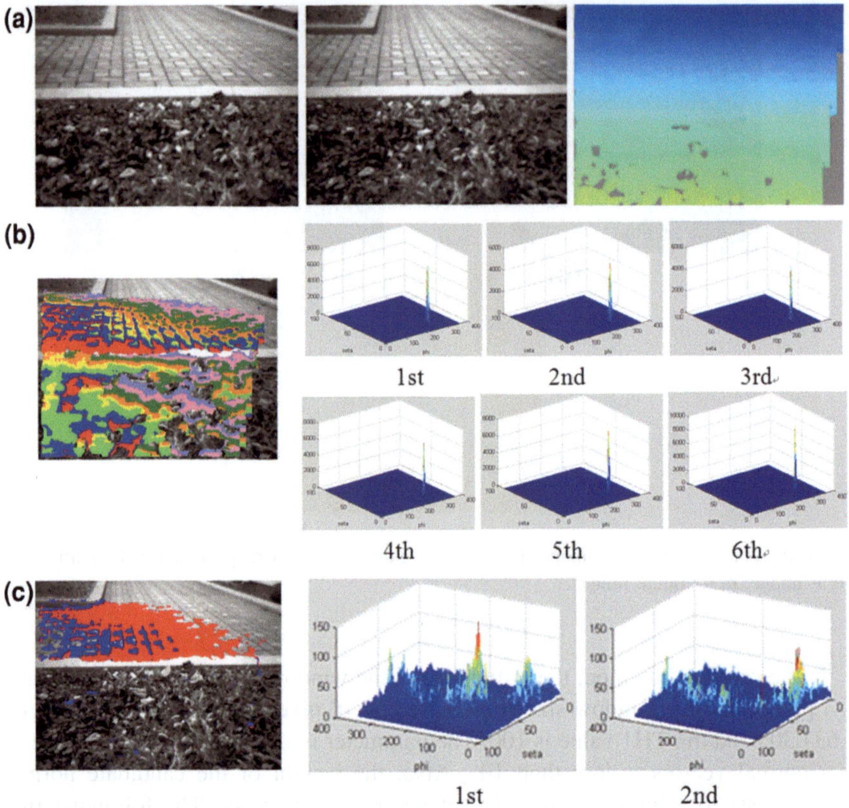

Fig. 11.9 Comparison experiments for an outdoor planar surface. **a** Original left, right image, and range image obtained from the stereo sensor; **b** Plane segment result of conventional RHT and peaks in Hough parameter space, in which each peak presents a plane; **c** Plane segment result of proposed RHT and peaks in parameter space

plane due to the effect of noises. While the conventional RHT method detects nine planes on an uneven flat and sidewalk flat surface. In the proposed method, the sampled data on an uneven flat surface is filtered by planarity evaluation while the conventional RHT method votes the parameters of sampled data on an uneven surface into the Hough parameter space, even though it is not a plane.

Figure 11.10 and Table 11.3 shows the experiment where there is an obstacle in the heading direction of the robot. The proposed method detected two planes of a flat surface and obstacle, while the conventional RHT method detected nine planes because of the influence of noises. Figure 11.10 and Table 11.3 don't include the parameter values of the 7–9th planes, because these are not the meaningful planes. The range data of stereo camera is easily affected by many external factors such as lighting, shape, texture, and so on. In the proposed method, sampled data that includes errors is filtered by planarity evaluation, but in the conventional RHT

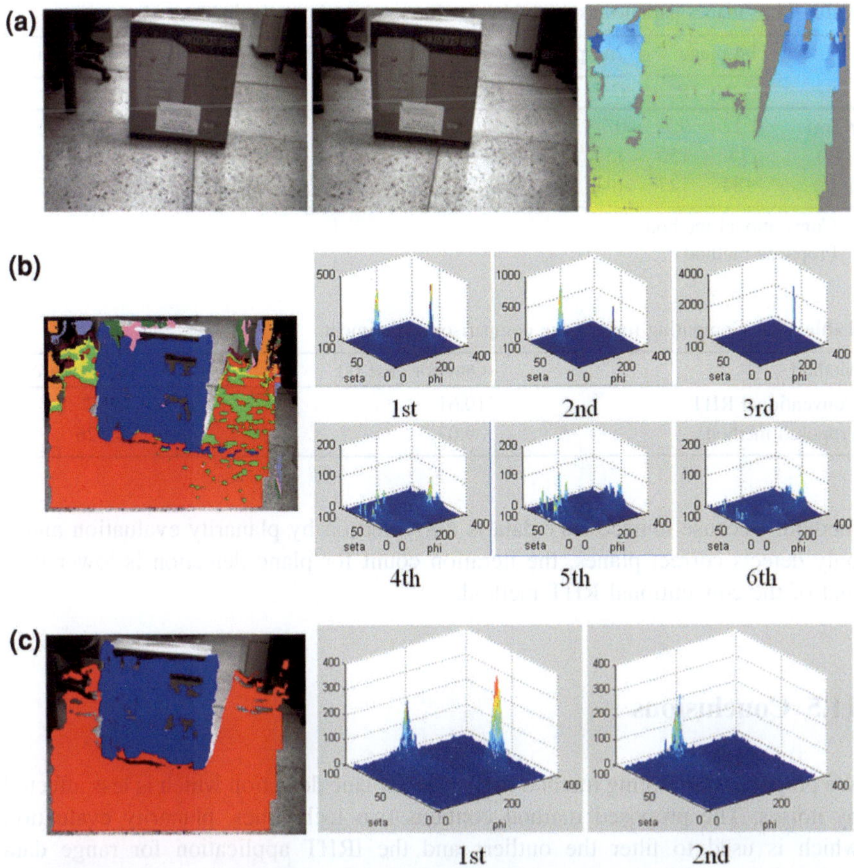

Fig. 11.10 Comparison experiments for an obstacle on the indoor surface

Table 11.2 Planes parameter of experiments for an outdoor planar surface

	1st plane		2nd plane		3rd plane		4th plane		5th plane		6th plane	
	C	P	C	P	C	P	C	P	C	P	C	P
ρ(m)	1.01	0.99	0.97	0.93	0.96	–	1.02	–	0.93	–	0.91	–
θ(°)	128	129	116	118	133	–	131	–	132	–	142	–

C Conventional method
P Proposed method

method, the parameters of the error data are not filtered and they make the wrong peak in Hough parameter space.

Table 11.4 shows the detection speed for two comparison experiments. The experiments prove that the proposed method is less affected by the noises and uneven flat surfaces, and it can detect planes faster than the conventional RHT

Table 11.3 Planes parameter of experiments for an obstacle on the indoor surface

	1st plane		2nd plane		3rd plane		4th plane		5th plane		6th plane	
	C	P	C	P	C	P	C	P	C	P	C	P
ρ (m)	1.22	0.7	0.7	1.25	1.25	–	1.26	–	0.57	–	1.02	–
θ (°)	113	150	154	113	113	–	113	–	133	–	117	–
φ (°)	−84	127	101	−87	−84	–	−83	–	36	–	−81	–

C Conventional method
P Proposed method

Table 11.4 Computing time of the comparison experiments

Method	Outdoor case (s)	Indoor case (s)
Conventional RHT	10.61	15.87
Proposed method	9.07	7.126

method. Because unnecessary data is not voted on by planarity evaluation and it only detects correct planes, the iteration count for plane detection is lower than that of the conventional RHT method.

11.5 Conclusions

We proposed a sampling method for accurate plane detection which is less affected by noises. The proposed method contains two techniques, planarity evaluation which is used to filter the outliers and the IRHT application for range data segmentation.

The planarity is tested by determining the angle of normal plane directions. These are determined by three randomly sampled vertices and the rotation points of the vertices. We can vote on accurate plane parameters, because most noises and nonplanar data are filtered from planarity evaluation. The scan window contributes to accurate plane detection and fast processing, because the planes have the local property, which means that the data of the same plane are located in a local region. The proposed method is verified by examples of a plane detection using real range data.

Acknowledgment This work was partly supported by the Ministry of Education, Science Technology (MEST) and Korea Institute for Advancement of Technology (KIAT) through the Human Resource Training Project for Regional Innovation, and Basic Science Research Program through the National Research Foundation of Korea (NRF) funded by the Ministry of Education, Science and Technology (2010-0027990) and the IT R&D program of MSIP/KEIT [Industry convergence original technology development projects, Development of context awareness monitoring and search system based on high definition multi-video].

References

1. Okada, K., Kagami, S., Inaba, J., Inoue, H.: Plane segment finder: algorithm, implementation and applications. In: IEEE International Conference Robotics Automation, pp. 2051–2058 (2001)
2. Illingworth, J., Kittler, J.: A survey of the Hough transform. Comput. Vision Graphics Image Process. 44 (1), 87–116 (1988)
3. Ben-Tzvi, D., Sandler, M.B.: A combinatorial Hough transform. Pattern Recogn. Lett. **11**, 167–174 (1990)
4. Xu, L., Oja, E.: Randomized Hough transform (RHT): basic mechanisms, algorithms, and computational complexities. CVGIP: Image Underst. **57**, 131–154 (1993)
5. Xu, L., Oja, E., Kultanen, P.: A new curve detection method: randomized Hough transform (RHT). Pattern Recogn. Lett. **11**, 331–338 (1990)
6. Kiryati, N., Eldar, Y., Bruckstein, A.M.: Probabilistic Hough transform. Pattern Recogn. Lett. **24**, 303–316 (1991)
7. Leavers, V.F.: The dynamic generalized Hough transform: its relationship to the probabilistic Hough transforms and an application to the concurrent detection of circles and ellipses. CVGIP: Image Underst. **56**, 381–398 (1992)
8. Kang, D.J., Lim, S.J., Ha, J.E., Jeong, M.H.: A detection cell using multiple points of a rotating triangle to find local planar regions from stereo depth data. Pattern Recog. Lett. **30**, 486–493 (2009)
9. Kälviäinen, H., Hirvonen, P., Xu, L., Oja, E.: Comparisons of probabilistic and non-probabilistic Hough transforms. In: Proceedings 3rd European Conference on Computer Vision, pp. 351–360 (1990)
10. Viola, P., Jones, M.: Rapid object detection using a boosted cascade of simple features. In: Proceedings of the IEEE Conference on Computer Vision and Pattern Recognition (2001)
11. Lu, W., Tan, J.: Detection of incomplete ellipse in images with strong noise by iterative randomized Hough transform (IRHT). Pattern Recogn. **41**, 1268–1279 (2008)
12. Point Grey company page. http://www.ptgrey.com

Chapter 12
Computer Vision Analysis of a Melting Interface Problem with Natural Convection

Gisele Maria R. Vieira, Fabiana R. Leta, Pedro B. Costa, Sergio L. Braga and Dominique Gobin

Abstract This study presents some Computer Vision techniques to analyze a phase change problem with natural convection. The analysis and interpretation of images are important to understand the phenomenon under study. Methods of image processing and analysis are used to validate the mathematical model and to automate the process of extracting information from the experimental model. The images produced by the experiment show the melting of a vertical ice layer into a heated rectangular cavity in the presence of natural convection and maximum density.

Keywords Computer vision analysis · Natural convection · Melting interface · Image segmentation · Digital filter

---

G. M. R. Vieira (✉)
Mechanical Engineering Department, Federal Center of Technological Education Celso Suckow da Fonseca—CEFET/RJ, Rio de Janeiro, RJ 20271-110, Brazil
e-mail: gisele.cefetrj@gmail.com

F. R. Leta
Mechanical Engineering Department, Universidade Federal Fluminense—UFF, Niterói, RJ 24210-240, Brazil
e-mail: fabiana@ic.uff.br

P. B. Costa
National Institute of Metrology, Quality and Technology, Av. Nossa Senhora das Graças, 50-Xerém, Duque de Caxias, RJ, Brazil
e-mail: pedrobcostarj@gmail.com

S. L. Braga
Mechanical Engineering Department, Catholic University of Rio de Janeiro—PUC-RJ, Rio de Janeiro, RJ 22453-900, Brazil
e-mail: slbraga@puc-rio.br

D. Gobin
FAST—CNRS—Université Paris VI, Campus Universitaire, Bât.502, Orsay 91405, France
e-mail: gobin@fast.u-psud.fr

F. R. Leta (ed.), *Visual Computing*, Augmented Vision and Reality 4,
DOI: 10.1007/978-3-642-55131-4_12, © Springer-Verlag Berlin Heidelberg 2014

List of nomenclature

A	Acrylic wall
C	Copper wall
c_p	Specific heat
$c^*(z^*,t^*)$, c(z,t)	Dimensional and dimensionless positions of the interface
E	Heat exchanger
Fo	Fourier number
$\mathrm{Gr}_{\mathrm{mod}}$	Modified Grashof number
H	Height of the enclosure
I	Insulation
k	Thermal conductivity
L	Liquid cavity maximum width
L_F	Latent heat
P^* and P	Dimensional and dimensionless pressures
Pr	Prandtl number
$\partial c/\partial \tau$	Velocity of the interface in the $\bar{n}$ direction
$\mathrm{R_i}$	Resistances
Ste	Stefan number
t^* and t	Dimensional and dimensionless times
T_{av}	Average temperature
T_{Fus}	Fusion temperature of the material
T_H and T_0	Temperatures of the hot and the cold walls
$\mathrm{T_i}$	Thermocouples
T_M	Temperature of the maximum density
$\bar{V}$	Dimensionless velocity vector
W	Removable window
y^* and y	Horizontal dimensional and dimensionless coordinates
z^* and z	Vertical dimensional and dimensionless coordinates
Z and Y	Computational dimensionless coordinates
α	Thermal diffusivity
$\Delta T = T_H - T_{\mathrm{Fus}}$	Temperature difference
ΔT_{max}	Maximum temperature interval considered
γ	Phenomenological coefficient
υ	Kinematic viscosity
θ	Dimensionless temperature
ρ_M	Maximum density
ρ_{ref}	Reference density

12.1 Introduction

Currently, digital images have widespread use in many applications. The increase in its use is mostly because much of the information that humans get from the world is by observing images, whether in their daily life, walking down the street, watching television, reading books, or in professional and scientific applications. In the last case, there are many data obtained from the analysis of photographs, spectrograms, thermal imaging, etc.

Computer Vision employs artificial systems that can extract information from digital images. It involves image acquisition, processing, and analysis.

As presented, Computer Vision techniques can offer support in many areas [1–4]. Images produced by experimental tests can be better understood through these techniques. In this work, image-processing techniques will be applied to better understand the behavior of a phase-changing material in the presence of natural convection. The analysis of this kind of problem has received increasing research attention and it is important for energy storage systems, thermal environment control, crystal growth processes, and other engineering applications. This study deals with the melting of a pure substance in the presence of natural convection in a rectangular enclosure due to a horizontal thermal gradient.

Several numerical and experimental studies have been reported in the literature concerning the problem of melting or solidification in the presence of convection. Examples of experimental studies can be found in the works of Wolff and Viskanta [5] and Bénard et al. [6]. In numerical analyses, the works of Kim and Kaviany [7], based on the Finite Difference method, and the works of Sparrow et al. [8], which pioneers in solving phase change problems by the Finite Volumes method, can be mentioned. Gobin and Le Quéré [9] performed a comparison exercise about melting with natural convection. This work applies different numerical procedures and models to a simple phase change problem.

Some fluids exhibit maximum density near their freezing points. In such case, the problem becomes even more complex because the hypothesis that the density varies linearly with temperature cannot be applied. This phenomenon occurs with water near 4 °C at atmospheric pressure, which is a temperature often found in several technological applications and in nature. Numerical and experimental works involving maximum density and thermal natural convection can be found in the literature, such as in the work of Lin and Nansteel [10] and Bennacer et al. [11]. Braga and Viskanta [12] and Kowalewsky and Rebow [13] analyzed the effect of maximum density in water solidification in a rectangular cavity. Tsai et al. [14] presented a numerical work about the effect of maximum density on laminar flows in tubes with internal solidification, involving mixed convection. A simple model of water freezing in a differentially heated cavity is used by Yeoh et al. [15].

This work is motivated by the need to gain a more complete understanding of the heat transfer process during the solid–liquid phase change that occurs with natural convection and maximum density.

12.2 The Phase Change Problem

This study deals with melting of a vertical ice slab upon a gravitational field in a rectangular enclosure due to a horizontal thermal gradient. The process is driven by thermally induced natural convection in the liquid phase.

12.2.1 Mathematical Model

The physical problem at time $t^* = 0$ is shown in Fig. 12.1. The asterisk superscript (*) is used to indicate the dimensional variables. In the initial condition considered half of the material volume is in the solid state, while the other half is in the liquid state. Initially, all the volume of the testing substance is set at its fusion temperature, i.e., $T_0 = T_{\text{Fus}}$. The melting process begins when the temperature of one of the vertical walls of the rectangular cavity, represented by T_H, is increased. H is the cavity's height and L is the width. The right vertical wall is kept isothermal at T_0 and the horizontal walls are adiabatic. The entire process is controlled by the natural convection in the liquid phase. The position of the interface at time t^* and level z^* is defined by its distance from de hot wall $c^*(z^*, t^*)$.

The hypotheses below were assumed in order to formulate the equations that manage the problem:

- The flow is laminar and two-dimensional.
- The liquid material is Newtonian and incompressible.
- The fluid's physical properties are constant, except for density in the buoyancy force term.
- The viscous dissipation is negligible.
- The density change of the material upon melting is neglected.
- It is assumed that the velocity of propagation of the melting front is several orders of magnitude smaller than the fluid velocities in the boundary layers on the vertical walls. This suggests that it is possible to divide the process in a number of quasi-static steps, separating, therefore, the melting front motion calculations from the natural convective calculations.

The coordinate system adopted, the time and the melting front were made dimensionless in the following way:

$$y = y^*/H; \quad z = z^*/H \tag{12.1}$$

$$t = t^* \upsilon / H^2; \quad c(z, t) = c^*(z^*, t^*)/H \tag{12.2}$$

Dirichlet thermal boundary conditions are taken on the vertical wall and at the interface, and the horizontal walls are adiabatic. In the liquid cavity, zero velocity dynamic boundary conditions are considered at the four walls.

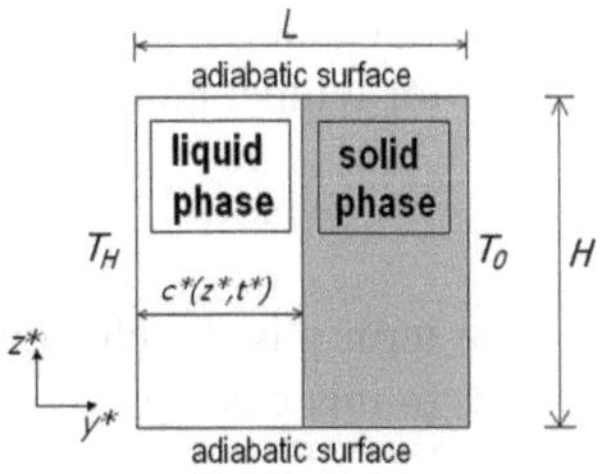

Fig. 12.1 Physical problem at time $t^* = 0$

Based on the previous hypothesis, the governing equations used in the non-rectangular liquid domain can be written in their dimensionless form as following:

$$\nabla \bullet \bar{\mathbf{V}} = 0 \tag{12.3}$$

$$(\bar{\mathbf{V}} \bullet \nabla)\bar{\mathbf{V}} = \nabla^2\bar{\mathbf{V}} - \nabla \mathrm{P} + \left(\frac{\mathrm{gH}^3}{\upsilon^2}\right)\left(\frac{\rho(T) - \rho_{\mathrm{ref}}}{\rho_{\mathrm{ref}}}\right)\bar{k} \tag{12.4}$$

$$(\bar{\mathbf{V}} \bullet \nabla)\theta = \frac{1}{\mathrm{Pr}}(\nabla^2\theta) \tag{12.5}$$

The dimensionless velocity vector $\bar{V}$ is given by: $\bar{V} = (\bar{V}^*\mathrm{H})/\upsilon$, where υ is the kinematic viscosity. The dimensionless temperature is given by $\theta = (T - T_{\mathrm{av}})/\Delta T$, where T_{av} is the average temperature given by $T_{\mathrm{av}} = (T_H + T_{\mathrm{Fus}})/2$. T is the dimensional temperature, and $\Delta T = T_H - T_{\mathrm{Fus}}$. The Prandtl number is given by $\mathrm{Pr} = \upsilon/\alpha$, and α is the thermal diffusivity. P is the dimensionless pressure, given by $P = (P^*\upsilon)/H^2$; ρ_{ref} is the reference density (equal to the average density of the interval imposed by the temperature of the walls) and $\bar{k}$ is the unitary vector in the vertical direction. At the moving interface, the energy balance equation is given by:

$$\nabla\theta \bullet \bar{\mathrm{n}} = \frac{\partial c}{\partial \tau} \tag{12.6}$$

The term $(\partial c/\partial \tau)$represents the local velocity of the melting front along the vector $\bar{\mathrm{n}}$, normal to the interface and $\tau = \mathrm{Ste} \times \mathrm{Fo}$, with Stefan number given by $\mathrm{Ste} = (c_p\Delta T)/L_F$, where c_p is the specific heat and L_F the latent heat. Fo is the Fourier number.

12.2.1.1 Density Approximation in the Buoyancy Term

As mentioned earlier, for fluids that reach an extreme density value at a specific temperature, it is not suitable to assume the hypothesis that the density varies linearly with temperature. Contrary to the linear estimate that predicts a unicellular flow, the maximum density formulation predicts a bicellular flow. As for water, the following equation, proposed by Gebhart and Mollendorf [16], provides very good results for temperatures below 10 °C:

$$\rho(T) = \rho_{\mathrm{ref}}.(1 - \gamma.|T - T_{\mathrm{ref}}|^{q}) \tag{12.7}$$

The term γ is the phenomenological coefficient given by $\gamma = 8 \times 10^{-6}$ °C^{-2}; $q = 2$. In this case, ρ_{ref} is equal to the maximum density of the fluid, also called ρ_M; and T_{ref} is equal to the temperature of the maximum density, which is also called T_M. For water $T_M = 3.98$ °C. A modified Grashof number based on the cavity height and on $\Delta T_{\max}$ (maximum temperature interval considered) have been defined:

$$\mathrm{Gr}_{\mathrm{mod}} = \frac{g.H^3.\gamma.(\Delta T_{\max})^2}{\upsilon^2} \tag{12.8}$$

The effect of variation of ρ is approximately symmetrical to the maximum density. The relative density variation that causes the flow in each cell is directly linked to the intervals between T_M and the wall temperatures. Then:

$$\Delta T_1 = T_H - T_M \tag{12.9}$$

$$\Delta T_2 = T_M - T_0 \tag{12.10}$$

The maximum temperature interval considered is:

$$\Delta T_{\max} = \mathrm{Max}(\Delta T_1, \Delta T_2) \tag{12.11}$$

12.2.2 Numerical Procedure

The numerical method used in this work has been successfully compared to the results of other researchers for the case of materials that do not present a maximum density to a comparison exercise proposed by Gobin and Le Quéré [9]. The numerical simulation technique is based on the hypothesis that the melting process is a succession of quasi-stationary states. In order to map the irregular space occupied by the liquid into a rectangular computational space, the dimensionless coordinates were transformed. The curvilinear coordinate system adopted is given by:

$$Z = z; \quad Y = y/C(Z) \tag{12.12}$$

where $C(Z) = c^*(z^*)/L$ and L is the maximum width of the liquid cavity. Z and Y are the computational dimensionless coordinates. Other details of the coordinate transformation method are shown in the work of Vieira [17]. The transformed

Table 12.1 Case test

Case test	T_H (°C)	ΔT (°C)	ΔT_I(°C) (Eq. 12.9)	ΔT_{max}(°C) (Eq. 12.11)	Ste	Pr	Gr_{mod} (Eq. 12.8)
	12	12	8	8	0.15	10.71	1.49×10^7

Table 12.2 Thermophysical properties

Properties $T_H = 12°$ C; $T_{av} = 6°$ C			
Thermal conduct: k (W/m.K)	0.6	Density: ρ (kg/m^3)	999.90
Specific heat: c_p (J/kg.K)	4200	Latent heat: L_F (J/kg)	3.3×10^5
Kinematic viscosity: υ (m^2/s)	1.5×10^{-6}	Thermal diffusivity: α (m^2/s)	1.4×10^{-7}

equations are discretized on a computational domain using the hybrid differencing scheme [18]. The pressure-velocity coupling is solved through the SIMPLE algorithm. The solution of the discretized equations is obtained through the ADI procedure. The grid defined on the computational domain is spaced irregularly to obtain a better resolution of temperature and velocity gradients at the solid walls. Among the types of grid tested, the 42 × 42 grid was chosen to present the results, based on the optimal balance of precision and computational time. Table 12.1 present the test performed.

The acceleration of gravity g used was 10.0 m/s^2. The values of the physical properties used are described in Table 12.2. These data have been obtained from the work of Gebhart et al. [19], based on the average temperature of each temperature interval considered.

12.2.3 Experimental Procedure

The experiments were performed in the rectangular test section shown in Fig. 12.2. The inner dimensions of the test section were: 187 mm in height, 187 mm in width, and 200 mm in depth. The top, bottom, and back acrylic walls were 12 mm thick. The observation window (front wall) was constructed with a pair of acrylic sheets 12 mm thick with a 12 mm air gap between them to eliminate condensation. The two copper sidewalls were held at constant temperatures and were 3 mm thick. The copper surfaces were oxidized to avoid corrosion. The inner surfaces of the test section, except the front wall and a slit of the top wall, were painted black to minimize the light beam reflection for photographic observations. The sidewalls were maintained at different temperatures by electrical resistances and circulating fluids coming from thermostatically controlled reservoirs.

A schematic diagram of the experiment apparatus is shown in Fig. 12.3. In this figure, the copper wall is represented by (C), the acrylic wall by (A), the heat exchanger by (E), the insulation by (I), the thermocouples by (T_i), and the resistances by (R_i). The apparatus was insulated with styrofoam to minimize the heat

Fig. 12.2 Semi-assembled test section

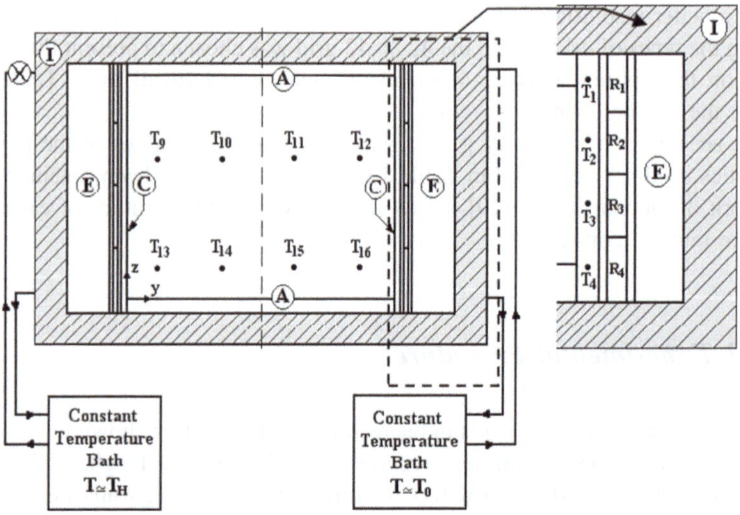

Fig. 12.3 Schematic diagram of the test section

gain from the environment to the experimental apparatus. During the visualization and photographic observations, part of the insulating material was removed. Distilled water was used as the testing fluid to avoid the presence of air bubbles. Four independent and controllable electrical resistances for each copper wall and two multipass copper heat exchangers were used. Omegatherm 201 conductive paste was used to ensure good thermal contact between the electrical resistances and the heat exchangers and the copper wall. The heat exchanger positioned on the side of the hot wall was connected through a valve system to the constant temperature bath. Four 0.5 mm thermocouples equally spaced in the z direction were embedded into each copper wall for continuous monitoring of their temperature. In this way, it was possible to maintain the temperature of the sidewalls uniform within ±0.1 °C of the desired temperature. In order to measure the temperature distribution inside the test section eight 1.0 mm thermocouples were inserted in the back wall through eight holes. They were arranged in two horizontal planes perpendicular to the front and back walls. In this way, two ranks of four evenly spaced thermocouples in the y direction measured the temperatures at two water depths in the z direction. This arrangement minimized heat conduction along the thermocouples and therefore reduced the measurement error. All thermocouples are type k (NiCr-NiAl) sheathed with stainless steel and were calibrated with an accuracy of ±0.1 °C. The thermocouples outputs were recorded by a data acquisition system at preselected time intervals between two consecutive measurements. These reservoirs, which are also called constant temperature baths, were connected to the coolant flow system by rubber tubes, which were insulated by foam pipes. Alcohol was used as the coolant fluid in the constant temperature baths.

The experimental case were performed with the cold wall maintained at 0 °C and the hot wall at $T_H = 12$ °C. This is an important interval to explore the maximum density phenomenon. The water was carefully siphoned into the test section to avoid introduction of air. To obtain the initial configuration of the solid phase of the fluid, it was necessary to freeze the substance with the test section turned 90° to the left before the beginning of the experiment (Fig. 12.4). In this way, the two constant temperature baths were set at low temperatures. After conclusion of the freezing process, the solid was heated up to a temperature close to its fusion point. In this way, the control point temperature was gradually increased and the temperature in the solid was carefully monitored. This heating regime was continued until all the system reached 0 °C. At this moment, the rest of the test section was filled with water at 0 °C. The desired water temperature (0 °C) was obtained by mixing cold water with crushed ice and allowing sufficient time for the equilibrium. The water was carefully siphoned into the test section to avoid introduction of air. The test section was turned 90° back to its original position, as shown by Fig. 12.4. Next, the valve attached to the hot wall was closed and the temperature of the bath connected to the valve was increased. When the temperature bath attained the desired temperature, the valve was reopened and the left wall was heated to T_H within 1 and 4 min depending on the T_H value. Once the initial conditions were attained, the melting run was started. Thermocouple data were collected at 1-min intervals. The temperature of the sidewalls was maintained constant throughout the data run with the help of the electrical resistances. Since the solid phase occupies a greater volume than the liquid phase, a feed line connected to the top wall was used to fill the test section with water at 0 °C.

To visualize the flow patterns, the water was seeded with a small amount of pliolite particles. Pliolite is a solid white resin with a specific gravity of 1.05 g/cm^3 that is insoluble in water. These tracer particles of small diameter (<53 μm) are neutrally buoyant. A beam from a 5.0 mW helium-neon laser was used as light source. The laser beam passed through a cylindrical glass rod to produce a sheet of laser light before passing through the test section wall. Photographs of the flow

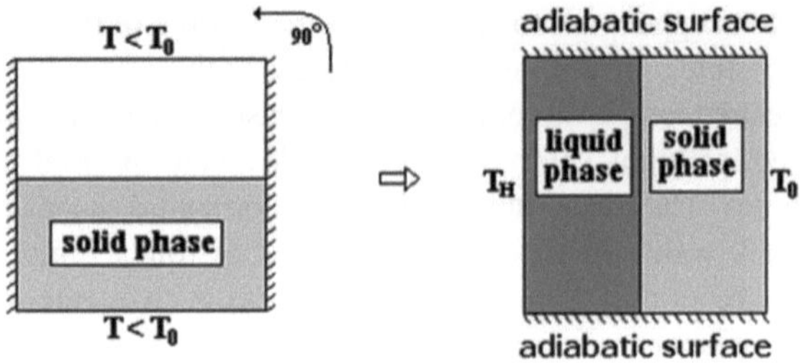

Fig. 12.4 Initial condition

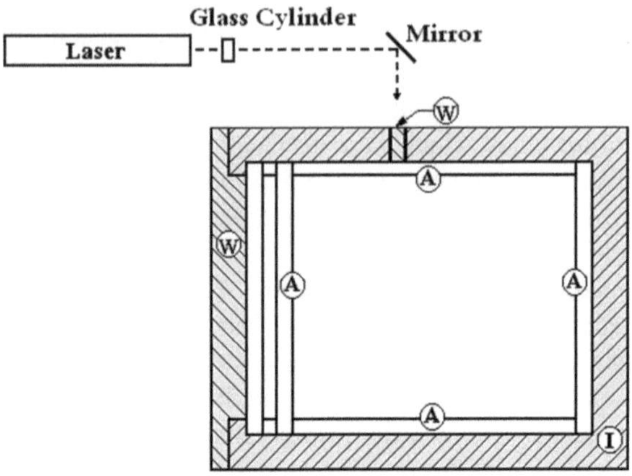

Fig. 12.5 The illumination system

patterns were taken using ASA 100 film (T-Max) and a 35 mm camera. The exposure time was about 60 s, with f = 5.6. Figure 12.5 shows schematically the illumination system. In this figure, the acrylic wall is represented by (A), the removable window by (W) and the insulation by (I).

12.3 Computer Vision Techniques

In order to determine the ice and water volume using the photographs obtained in the experimental procedure, it is necessary to apply a segmentation technique in these images. The target is the segmentation of ice and water areas into two different pixel groups. This process can be so easy to many people, in Fig. 12.6, for instance the separation of ice and water by human vision is very intuitive. However, to perform the segmentation process automatically using Computer Vision techniques it should be well accomplished.

A digital image can be considered as an array of $N_1 \times N_2$ elements where $1 \leq n_i \leq N_i$. Therefore, for each position (n_1, n_2) we have a $f(n_1, n_2)$ function, which represents the intensity of each pixel illumination (Fig. 12.7). In grayscale images, this function can be $0 \leq f(n_1, n_2) \leq 255$.

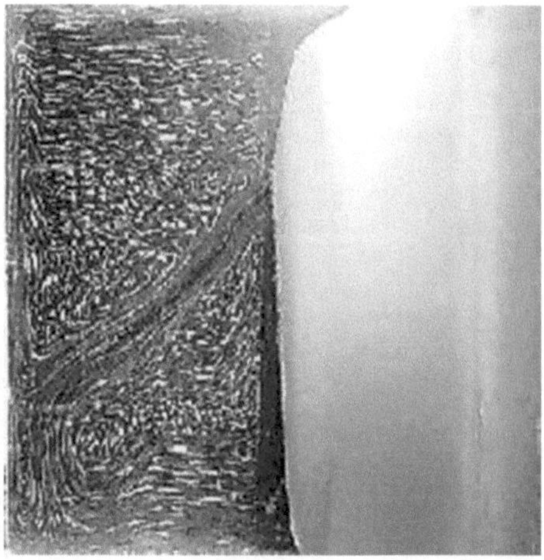

Fig. 12.6 Picture of the experiment

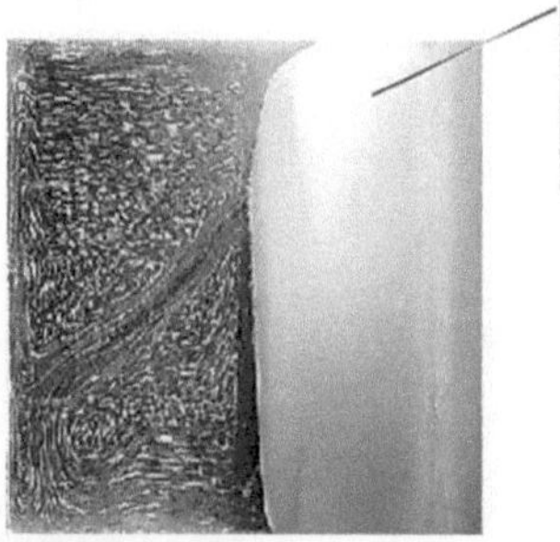

233	230	200	141	121
231	230	199	143	121
229	233	205	142	119

Fig. 12.7 Example of an image representation

Image segmentation is the process of partitioning an image into meaningful parts [2]. Segmentation techniques can be thresholding, boundary detection, or region growing.

In the thresholding process, generally it used the analysis of pixels grayscale frequency (histogram) to determine a threshold (*l*) parameter (Eq. 12.13). The image result is a new one *I*(*x*,*y*) [20].

$$I(x,y) = \begin{cases} 0, & f(x,y) < l \\ 255, & f(x,y) \geq l \end{cases} \tag{12.13}$$

This method can achieve good results in images with bimodal histograms (Fig. 12.8), different from the example image histogram shown in Fig. 12.9, where there is no two noticeable picks. In bimodal histograms case, it is simple to choice a threshold parameter *l* to segment the two different regions in an image. However, the method of amplitude threshold is not a good procedure in many cases, as can be seen in Fig. 12.9. The reason of this bad result is the fact that the luminance isn't a distinguishable feature for water and ice in this particular example.

Fig. 12.8 Bimodal histogram

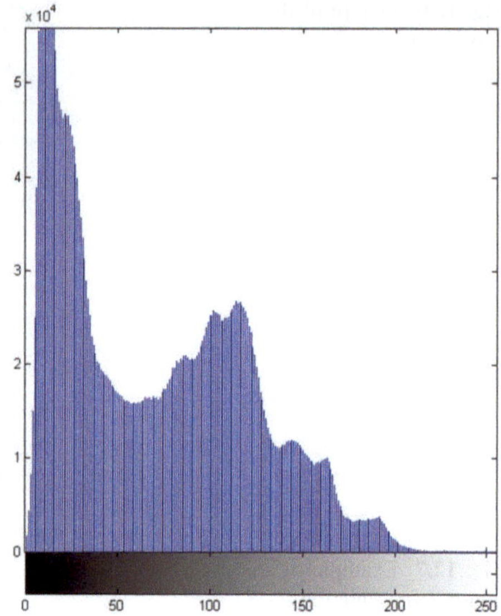

Fig. 12.9 Histogram of Fig. 12.6

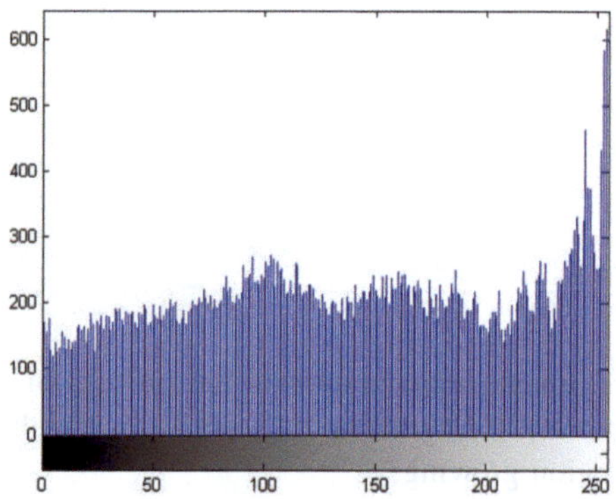

The result shown in Fig. 12.10 was obtained by the application of a segmentation technique that uses the smallest value between the two higher picks of the histogram (Fig. 12.9).

Due to uniformity of the brightness values around the ice in the image and the features inside the water, it was necessary to use a more accurate algorithm to determine the ice and water area in the experiment. Region growing method is not adequate because it forms distinguished regions by combining pixels with similar grayscale intensity. Observing the Fig. 12.6, it can be noted that many small regions could be highlighted. For that reasons, the chosen algorithm used to separate ice and water is a boundary enhancement method.

Fig. 12.10 Binary image

12.3.1 Boundary Enhancement Techniques

Boundary enhancement methods use information about the intensity differences between neighboring regions in order to separate these regions [2]. Boundaries or edges are discontinuities of a color or grayscale intensity in an image. These intensity discontinuities can help to identify the physical extensions of objects in a digital image. Figure 12.11 shows four possible schemas of an edge considering a continual one-dimensional domain [21].

The edges generally are characterized by its amplitude and slope angle. A detector algorithm must find the slope midpoint that characterizes the edge. This task would be very simple if the edge slope angle is 90°, called a step edge. However, this inclination is not found in digital images, step edges exist only for synthetic and graphical images. In a real picture, the transition is very irregular.

There are many different methods to enhance an edge in order to segment the different regions in an image. However, each one has different approaches that generate different results, depending on the specificity of the image. Figure 12.12 shows the obtained results considering the application of some edge detector methods (Log, Prewitt, Roberts, Sobel, Zero-Cross).

When different methods are used, many different aspects in images can be found. The algorithms of edge detection can also modify the results when these methods are used to measure objects, as it is the main objective of this chapter.

In a previous work, diverse edge detector methods were evaluated [22]. The aim of this work was to reach the better result in an aperture area calibration, considering edge detection. Thus, six methods for edge enhancement were tested and compared each one with average value of all results. This comparison was carried out through to define if some of the methods are not appropriate for aperture area measurement. We used the following edge enhancement algorithms: Sobel, Prewitt, Roberts, Laplacian, Zero-Crossing, and Canny [23, 24].

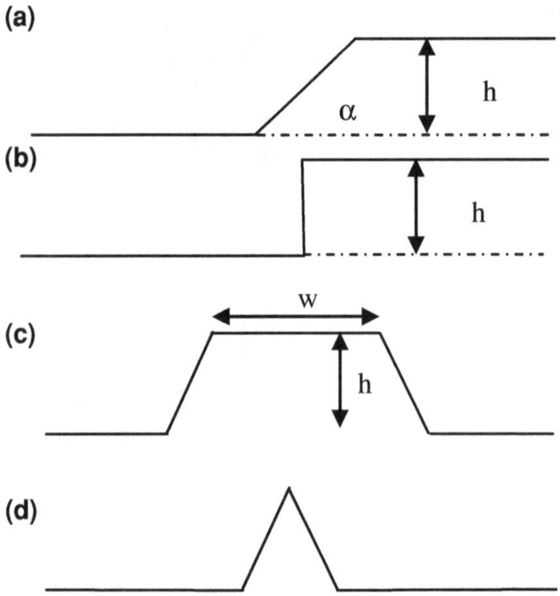

Fig. 12.11 Edges examples: **a** Ramp edge, **b** Step edge, **c** Line, **d** Roof edge

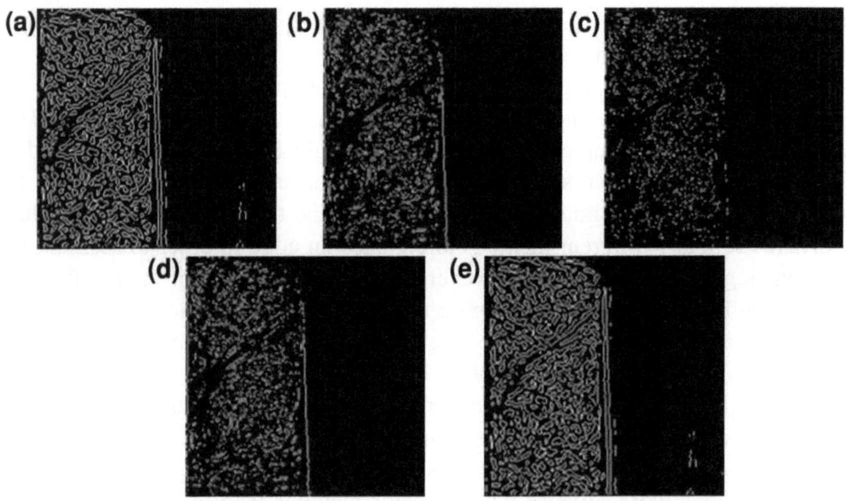

Fig. 12.12 Methods of edge detection: **a** Log, **b** Prewitt, **c** Roberts, **d** Sobel, **e** Zero-cross

The Canny algorithm was considered the best method. With the criteria used in this algorithm, even if the edge has a small slope and amplitude of intensity sufficiently reduced, the Canny method was shown efficient.

12.3.2 Canny Edge Detector

The objective of this experiment is to determine the area of ice and water in the images. With that purpose an algorithm of edge detection was used. Considering the discrete description of image, several techniques can be used for edge enhancement and detection in digital images. Here, the chosen technique was the Canny method.

John Canny, in 1986, in his paper "the computational Approach to edge detection" described a method for edge detection purpose [25]. In this paper, Canny treats the problem of edge detection establishing suitable criteria, so that one can achieve better results than other known methods.

The first criterion is the called *low error rate*, which considers that "edges occurrences can't be discarded and false edges can't be found." The second criterion refers to "the distance between the center of the edge and the points in its neighborhood that should be minimized." The last one was implemented because the first two were not sufficient to eliminate the possibility of multiple results for only one edge. Therefore, the third criterion considers that the algorithm should find only one result for a single edge.

The application of the Canny method is done by steps. The first step is the use of a Gaussian filter. The objective of this filter is to remove undesirable noises and enhance important features. This filter is accomplished by the convolution of the image, using a mask that is different depending on the specific filter.

The convolution operation consists in the application of a bidimensional filter taking into account the image grayscale intensity domain. Each pixel in the original image $x(n_1, n_2)$ is transformed in a new one $y(n_1, n_2)$ (Eq. 12.14).

$$y(n_1,n_2) = h(n_1,n_2) \otimes x(n_1,n_2) \; = \sum_{k_1=-\infty}^{\infty} \sum_{k_2=-\infty}^{\infty} x(k_1,k_2)h(n_1-k_1,n_2-k_2) \tag{12.14}$$

The term h represents the convolution mask of the used filter.

In order to implement the filtering of images through convolution, it is necessary to slide a mask (kernel) across the image (Fig. 12.13). The new pixel value $y(n_1, n_2)$ is the weighted sum of the input pixels $x(n_1, n_2)$ within the mask where the weights are the values of the filter assigned to every pixel of the window itself [21].

The choice of convolution mask is done according to the filter and the resolution desired. If the mask is large, the effect of the algorithm will be less observed. On the other hand, when the mask is small we will spend more computational processing.

For the Gaussian filter, its mask is obtained by a discreet Gaussian fit. The Gaussian curve has the characteristic of augment values that are close to the average value and to reduce the intensity values that are far from it. The Gaussian

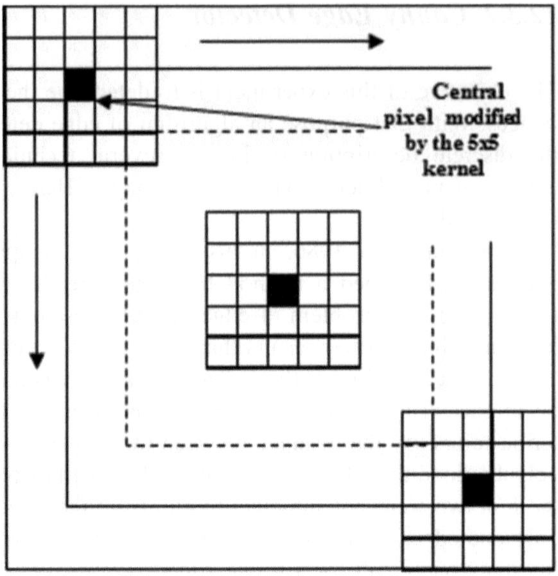

Fig. 12.13 Convolution procedure

curve width is measured considering the number of standard deviations.The Gaussian curve equation is given by:

$$g_\sigma(x) = \frac{1}{\sqrt{2\pi\sigma}} e^{\frac{-x^2}{2\sigma^2}} \tag{12.15}$$

To use this equation in image processing, it is necessary to generalize the Gaussian curve into a two-dimensional space.

$$G(x, y) = \frac{1}{2\pi\sigma_x\sigma_y} e^{\frac{-x^2}{2\sigma_x^2}+\frac{-y^2}{2\sigma_y^2}} \tag{12.16}$$

where σ is the standard deviation and the points (*x*, *y*) correspond to the distance between the coordinates and the mask center. For the application of this chapter, we used a 7 × 7 dimension mask with a standard deviation equal to $\sigma = 1, 4$.

$$\frac{1}{115} \cdot \begin{bmatrix} 2 & 4 & 5 & 4 & 2 \\ 4 & 9 & 12 & 9 & 4 \\ 5 & 12 & 15 & 12 & 5 \\ 4 & 9 & 12 & 9 & 4 \\ 2 & 4 & 5 & 4 & 2 \end{bmatrix} \tag{12.17}$$

The second step of the algorithm consists in determining the edge magnitude from the gradient of the image. It can be found using a Sobel operator. This method uses two 3 × 3 masks that approximate the derivatives of the image, for direction x and direction y.

$$\begin{bmatrix} -1 & 0 & 1 \\ -2 & 0 & 2 \\ -1 & 0 & 1 \end{bmatrix} \begin{bmatrix} 1 & 2 & 1 \\ 0 & 0 & 0 \\ -1 & -2 & -1 \end{bmatrix} \tag{12.18}$$
$$Gx \qquad\qquad Gy$$

Using the obtained results from both directions, the gradient of the image can be calculated (Eq. 12.19).

$$G = \sqrt{G_x^2 + G_y^2} \tag{12.19}$$

The next step is to determine the edge directions in the image. Using the values of the gradient and Eq. 12.20, we can determine these directions.

$$\theta = a\tan\left(\frac{G_y}{G_x}\right) \tag{12.20}$$

When the algorithm is implemented we must consider the cases where Gx is equal to 0. In this case, we will generate a mathematical error. Therefore, in these points the algorithm must analyze the gradient in Gy direction in order to decide which is the edge direction (0°, 45°, 90°, 135°).

Then the algorithm realizes a suppression process. This process consists in analyze, using a 3 × 3 matrix, all the points in all obtained directions calculating their gradient angle. We will consider as an edge only the points that have higher grayscale intensity than their correspondent neighbors in the opposite directions.

Finally, to find the edges, the image is analyzed using two threshold points, a high and a low one. When a pixel is greater than the high threshold value, its neighbor pixels need to have higher values than the low threshold point, in order to generate a continuous edge.

12.3.3 Image Processing Results

The Canny method was applied in five images of the experiment presented in Sect. 2.2. To reduce the possibilities of error in this algorithm and to obtain more efficient computational results, the images had been divided into two parts, because in this experiment the space occupied by the ice wasn't higher than 50 % of the total reservoir area (Fig. 12.14).

The results of Canny method in the images are shown in Fig. 12.15.

Applying the algorithm, it is possible to separate completely water and ice in each image. These images are used as masks in the original images (Fig. 12.16).

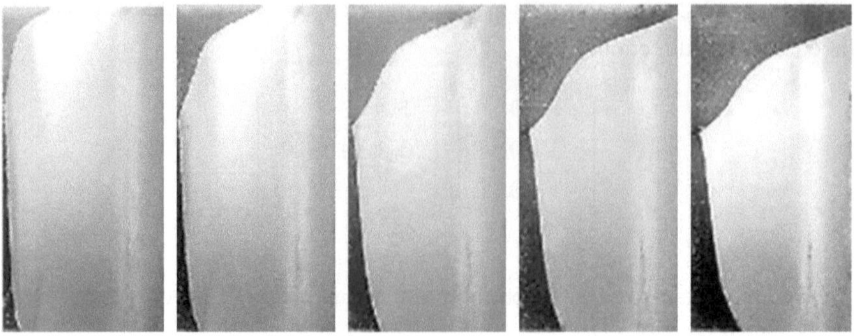

Fig. 12.14 Images cropped to apply the Canny method

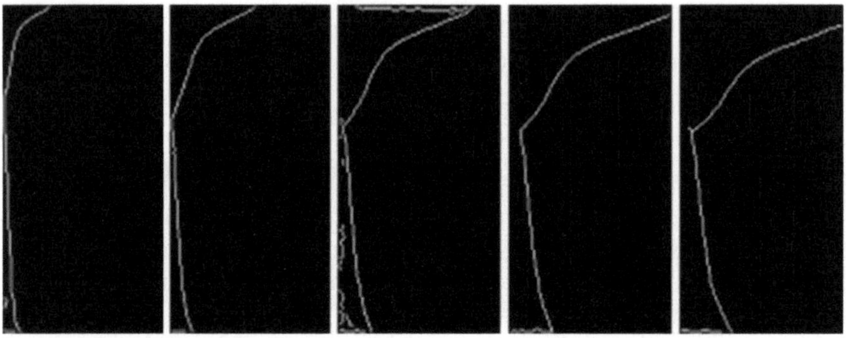

Fig. 12.15 Results of Canny method

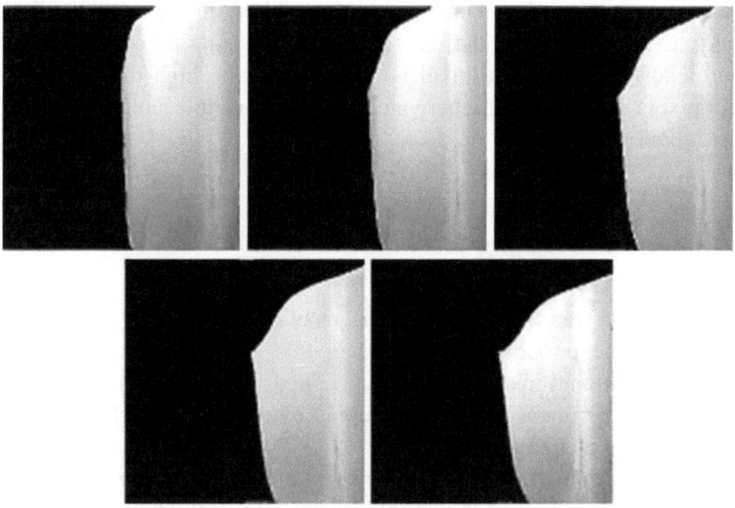

Fig. 12.16 Ice and water areas

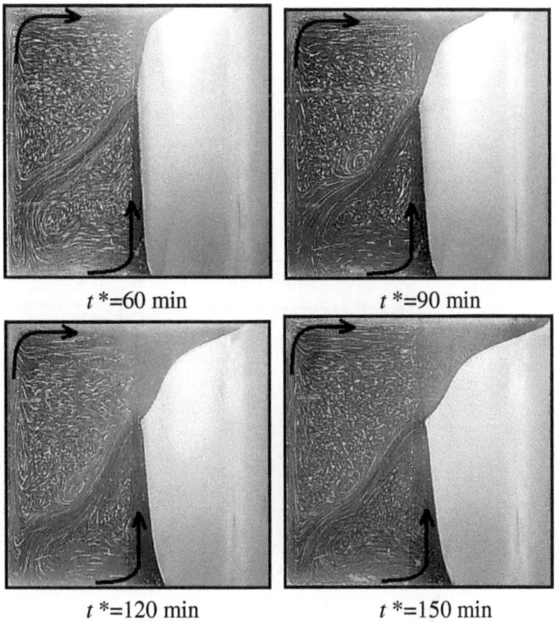

Fig. 12.17 Photographs of the evolution of the melting front produced by the experimental procedure

The final result consists in the distinguished ice (white) and water (black) pixels in each image.

After that it is possible to determine ice and water area, performing the sum of the white and black pixels, respectively. Considering the reservoir dimensions and the picture resolution, we calculate the volume of the solid and liquid state.

12.4 Results and Discussions

The photographs of the evolution of the melting front produced by the experimental procedure are shown in Fig. 12.17, and the isotherms and the evolution of the melting front produced by the numerical procedure are shown in Fig. 12.18.

The experimental and numerical tests show that the influence of the maximum density is evident. The counterclockwise cell is close to the cold wall and the clockwise cell is close to the hot wall. As the isothermal values of the dominant cell are greater than T_M, the melting front advances faster at the top of the cavity. The incidence of the heated fluid normal to the interface is responsible for the accelerated melting of ice at this point. These figures show a good qualitative agreement on the evolution of shape of the melting front.

The evolution of the liquid fraction (melted fraction) of the experiment is shown in Fig. 12.19. The melted fraction is the ratio of the liquid volume to the total volume of the enclosure. In the beginning of the process there were 50 % of

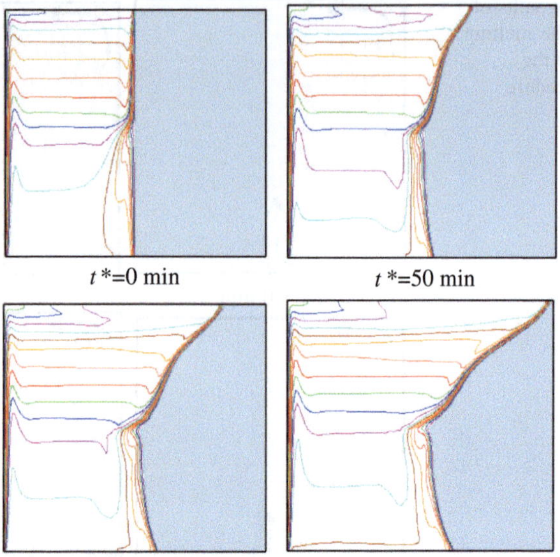

Fig. 12.18 Isotherms and the evolution of the melting front produced by the numerical procedure

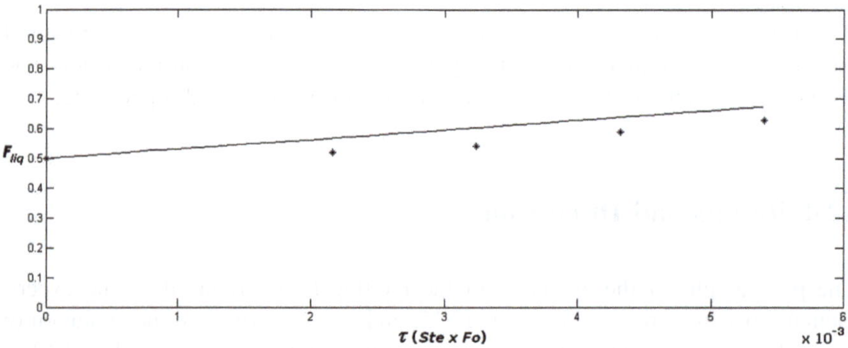

Fig. 12.19 Liquid fraction

liquid in the cavity, as the initial condition established previously. The results show a linear evolution of the melted fraction. There is a good agreement among the numeric and experimental results, obtained by the images processing, although it can be noted that the experimental points are clearly below the curve obtained by the numeric procedure. That is, for the same time, the melting front in the numeric simulation moves faster than for the experiments.

12.5 Conclusions

The most significant changes and improvements in scientific measurements are due to the development of measuring methods with higher accuracy results and lower uncertainties. Most researchers have developed and improved methods and systems of noncontact measurement, where all steps in the procedure use Computer Vision techniques [22].

Following this trend, we proposed to evaluate some Computer Vision techniques to support the experimental analysis of a melting problem with natural convection. The complete knowledge of experimental results is very important for the understanding of the complex phenomenon that involves moving boundary problems. Systems that can extract information from digital images help in the validation of numerical and experimental results, and provide a better understanding of the phenomenon under study. The results of this study can be applied in energy storage systems and other engineering applications. The experiments consider two water volumes in the solid and liquid states. The objective is to analyze the behavior of the moving interface and the evolution of the liquid fraction.

The problem was solved using image segmentation, which is the most studied topic in image analysis. Considering that there is not a universal method to segment all images [2], the aim of this study is unique.

The results using Computer Vision techniques to extract the ice and water volume obtained during the experimental procedure matched the values obtained in the numerical modeling. Given the uncertainties of the measurement process and experimental procedure, the final results are considered very satisfactory.

The segmentation algorithm used gave a better definition for the edges, reducing the number of incorrect values in the enhancement of the ice-water interface.

References

1. Bovik, A. (ed.): Handbook of Image and Video, 2nd edn. Elsevier Academic Press, New York (2005)
2. Goshtasby, A.A.: 2-D and 3-D Image Registration. Wiley, Hoboken (2005)
3. Palmer, S.E.: Vision Science—Photons to Phenomenology. The MIT Press, Cambridge (1999)
4. Nielsen, F.: Visual Computing: Geometry, Graphics and Vision. Charles River Media Inc., Massachusetts (2005)
5. Wolff, F., Viskanta, R.: Melting of a pure metal from a vertical wall. Exp. Heat Transf. **1**, 17–30 (1987)
6. Benard, C., Gobin, D., Martinez, F.: Melting in rectangular enclosures: experiments and numerical simulations. J. Heat Transf. **107**, 794–803 (1985)
7. Kim, C.J., Kaviany, M.: A numerical method for phase-change problems with convection and diffusion. Int. J. Heat Mass Transf. **35**, 457–467 (1992)
8. Sparrow, E.M., Patankar, S.V., Ramadhyani, S.: Analysis of melting in the presence of natural convection in the melt region. J. Heat Transf. **99**, 520–526 (1977)

9. Gobin, D., Le Quéré, P.: Melting from an isothermal vertical wall. Comput. Assist. Mech. Eng. Sci. **7–3**, 289–306 (2000)
10. Lin, D.S., Nansteel, M.W.: Natural convection heat transfer in a square enclosure containing water near its density maximum. Int. J. Heat Mass Transf. **30**, 2319–2329 (1987)
11. Bennacer, R., Sun, L.Y., Toguyeni, Y., et al.: Structure d'écoulement et transfert de chaleur par convection naturelle au voisinage du maximum de densité. Int. J. Heat Mass Transf. **36–13**, 3329–3342 (1993)
12. Braga, S.L., Viskanta, R.: Transient natural convection of water near its density extremum in a rectangular cavity. Int. J. Heat Mass Transf. **35–4**, 861–887 (1992)
13. Kowalewsky, T.A., Rebow, M.: Freezing of water in a differentially heated cubic cavity. Int. J. Comput. Fluid Dyn **11**, 193–210 (1999)
14. Tsai, C.W., Yang, S.J., Hwang, G.J.: Maximum density effect on laminar water pipe flow solidification. Int. J. Heat Mass Transf. **41**, 4251–4257 (1998)
15. Yeoh, G.H., Behnia, M., de Vahl Davis, G., et al.: A numerical study of three-dimensional natural convection during freezing of water. Int. J. Num. Method Eng **30**, 899–914 (1990)
16. Gebhart, B., Mollendorf, J.: A new density relation for pure and saline water. Deep Sea Res. **24**, 831–848 (1977)
17. Vieira, G.: Análise numérico-experimental do processo de fusão de substâncias apesentando um máximo de densidade. Ph.D. thesis, Pontifícia Universidade Católica do Rio de Janeiro, Rio de Janeiro, Brazil (1998)
18. Patankar, S.V.: Numerical heat transfer and fluid flow. Hemisphere, McGraw-Hill, New York (1980)
19. Gebhart, B., Jaluria, Y., Mahajan, R.L., et al.: Buoyancy-induced flows and transport. Hemisphere Publishing Corporation, New York (1988)
20. Ritter, G.X., Wilson, J.N.: Handbook of Computer Vision Algorithms in Image Algebra. CRC Press, Florida (1996)
21. Pratt, W.K.: Digital Image Processing, 4th edn. Willey, Canada (2007)
22. Costa, P., Leta, F.R.: Measurement of the aperture area: an edge enhancement algorithms comparison. In: Proceeding of IWSSIP 2010—17th International Conference on Systems, Signals and Image Processing, pp. 499–503, Rio de Janeiro, Brazil (2010)
23. Conci, A., Azevedo, E., Leta, F.R.: Computação Gráfica–Teoria e Prática [v.2]. Elsevier, Rio de Janeiro (2008)
24. Canny, J.: A computational approach to edge detection. IEEE Trans. Pattern Anal. Mach. Intell. PAMI **8**(6), 679–698 (1986)
25. Madisetti, V., Williams, D.B. (eds.): Digital Signal Processing Fundamentals. CRC, USA (1998)

Chapter 13
Real-Time Visualization and Geometry Reconstruction of Large Oil and Gas Boreholes Based on Caliper Database

Fabiana R. Leta, Esteban Clua, Diego C. Barboza, João Gabriel Felipe M. Gazolla, Mauro Biondi and Maria S do Souza

Abstract The evaluation of technical and economical viability before starting the drilling process of a gas or oil reserve is very important and strategic. Among other attributes, the soil structure around the borehole must be analyzed in order to minimize the risks of a collapse. This stability analysis of a gas or oil reserve is a challenge for specialists in this area and a good result at this stage could bring a deep impact in reduction of drilling costs and security. A tool known as caliper [1] is inserted into the drilling spot to perform a series of measurements used to evaluate the well's viability. For each position along the borehole, information such as sensors' position, orientation, well's resistivity, and acoustic data are obtained and recorded. These data allow the user to find flaws in the soil, leaving to the geologist the decision whether the well is feasible or not, and help them to study possible actions to minimize its usage risk. Currently, the data obtained by the caliper are used for the visualization of individual sections of the well, projected in a bi-dimensional plane, considering a cylinder projection. However, an

F. R. Leta (✉) · M. Biondi · M. S do Souza
Mechanical Engineering Department, Computational and Dimensional Metrology Laboratory, Universidade Federal Fluminense—UFF, 24210-240 Niterói, RJ, Brazil
e-mail: fabianaleta@id.uff.br

M. Biondi
e-mail: mbiondi@lmdc.uff.br

E. Clua · D. C. Barboza · J. G. F. M. Gazolla
Computer Science Institute, Universidade Federal Fluminense—UFF, 24210-240 Niterói, RJ, Brazil
e-mail: esteban@ic.uff.br

D. C. Barboza
e-mail: dbarboza@ic.uff.br

J. G. F. M. Gazolla
e-mail: gazolla@ic.uff.br

F. R. Leta (ed.), *Visual Computing*, Augmented Vision and Reality 4,
DOI: 10.1007/978-3-642-55131-4_13,

overview of the borehole's entire structure is necessary for a higher quality analysis. This work proposes a novel technique for a precise geometry reconstruction of the borehole from these data, allowing the geologist to visualize the borehole, making easier to find possible critical points and allowing an intuitive visualization of a large set of associated data. The three-dimensional geometry reconstruction is made from data collected from the caliper log, which includes the tool orientation and sensors' measures for each section. These measures are used as control points for the construction of smooth layers, through splines interpolation [2]. Finally, the sections are joined in sequence to form a polygonal mesh that represents a reliable vision of the borehole in three dimensions.

Keywords Well bore model · Oil well visualization · Caliper log · Three-dimensional visualization · Closed natural cubic splines

13.1 Introduction

Scientific real-time visualization techniques are important tools to understand what is almost impossible to be seen in numbers. When this visualization approaches are built into an interactive environment, a powerful workflow is available for the production and analysis pipeline. In the oil and gas industry, it is always very important to evaluate the technical and economical feasibility before starting the drilling process of a gas or oil reserve. The soil structure around the well must be known, in order to minimize the risks of collapse. The stability analysis of a gas or oil reserve is a central issue and a challenge for specialists. Good results at this stage could bring a deep impact in reduction of drilling costs. In the later years, wells have become longer and with higher inclination making it even more challenger and expensive.

According to Aadnoy [3], for years people have estimated the time loss associated with unexpected borehole stability problems to account for 10–15 % of the time required to drill a well. Since the rig time is the major cost factor in drilling operation, we understand that borehole stability problems are very costly for the industry. Borehole collapse is possibly the most costly single problem encountered during drilling of a well, and there is not a trivial solution for the problem.

This theory states that drilling generates changes in the stress field of the formation due to supporting material losses, inducing stresses that can result in more trouble [4]. If the stress is higher than the rock strength, rocks can cause the borehole collapse. So the main fundamental of the solution proposed in this work is to possibly prevent and reduce instability problems.

The better the planning of a well, the greater are the chances of success to be achieved. In this context, the word success means achieving the objectives of the project according to the safety standards of the company, time, and costs consistent with the ones of the market. However, to achieve this success, Geologists,

Reservoir Engineers and Production Engineers should not only care to the optimization of the production of the reservoir but should also be aware of the operational risks and expenses of the project.

When the well is drilled and a part of its formation is removed, there is instability of stresses. In that moment comes one of the most important stages of the project of a well which is ensuring the stability of it, and to achieve this goal it is necessary to evaluate factors like: rock resistance, temperature variations, well trajectory, and geometry.

This work proposes a novel framework for a 3D geometric reconstruction and visualization of the bare holes and its components. Data structures and processes adopted allow a real-time interaction, using dedicated but not expensive graphic hardwares. New interfaces and data manipulation are proposed, increasing the workflow process. In order to maintain this real-time requirement, our system builds the geometry using a thread-based reconstruction model and level of detail algorithms for maintaining huge datasets in the memory.

13.2 Related Works

In [1] it presented a borehole visualization approach, more focused on stress and breakout analysis. Barton and Zoback [5] and Peska and Zoback [6] presents important factors to be considered in order to model the requirements associated to a detailed data processing and analysis of this kind of data, that will be considered on the proposed system.

In [7] the authors present a technique for modeling a 2D approach, so important requirements of these structures are maintained and can be analyzed, such as perforation fluid weight, rock strength, and hole geometry. Figure 13.1 shows the data obtained by a four arms caliper log. The 2D developed system shows an approximation of the borehole shape in a specific height (Fig. 13.2). Each colored profile is related to a caliper arm and is used for the 3D reconstruction.

Despite allowing the user to obtain various types of information, this tool does not present a global spatial view of the three-dimensional geometry of the borehole, even the most experienced specialists can experience difficulties to imagine its geometry, being a restriction, but a key point to the proposed three-dimensional tool.

Debogurski [8] presents a procedural terrain generation using the recent Marching Cubes Histogram Pyramids (also known as HPmarcher) implementation. Perlin Noise function is used to procedurally create the terrain. This is an important approach related to our proposal, since it is a GPU-based approach for the modeling optimization, which allows a huge number of polygons to be handling in a Real-Time System.

This work merges characteristics from most of these works, enhancing the visualization process, since important data may be mapped into a huge and massively geometry borehole reconstruction.

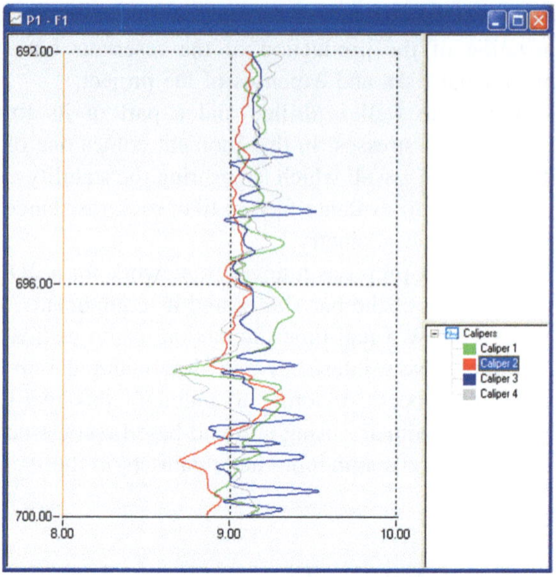

Fig. 13.1 Profiles obtained by a 4-arm caliper log

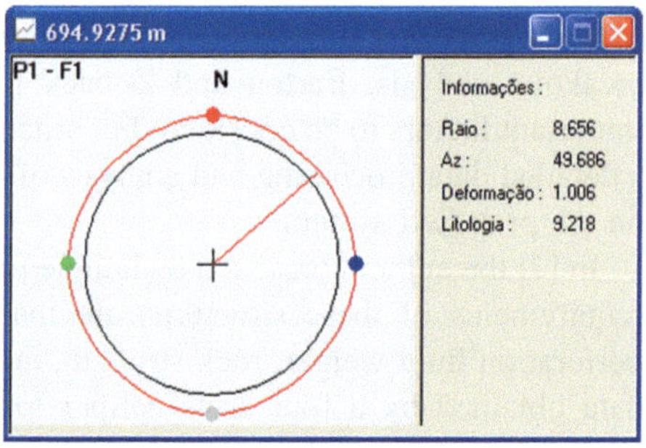

Fig. 13.2 A screen of the 2D caliper analysis system

13.3 Methodology

Our methodology is divided into six steps (Fig. 13.3). The process begins describing how the data is acquired by the Caliper tool. After this, the data is converted into a specific data structure. The third step consists on the calculations necessary for the geometric consistency, and on the fourth step some optimization are done, which means that we try to send to the next step only the necessary vertexes. To finish the process the user can make the analysis with the tool information and finally obtain useful knowledge. A more detailed explanation is described in the following subsections. This workflow is related to the usage process of the proposed system.

Fig. 13.3 The framework of the 3D Caliper tool (GeoPoço 3D)

13.3.1 Data Acquisition

A borehole failure can be detected by different measurement instruments, like the following caliper logs: Borehole Geometry Tool; Resistivity logs; Ultrasonic logs. The caliper tool is inserted into drilling spot to perform a series of measurements used to evaluate the borehole's feasibility [1]. It measures the borehole diameter variation by depth, i.e., its shape and can also give other important information, as well hole irregularities or damages. The caliper log can have two, four, or more extendable arms arranged around the tool. Figure 13.4 shows a simple schema of the tools and obtained data. Among all these measures P1AZ represents the alignment of the first arm with the magnetic north in degrees, DEVI represents the tool's inclination and combined with HAZI gives information about the direction. All this informations are strictly necessary to reconstruct the borehole geometry, and these variations on every section will allow the borehole to make curves, creating the sensation of a real borehole and not just a simple geometric form like a cylinder.

For each position along the borehole, information such as sensors' position and orientation, rock's resistivity and acoustic data are obtained. These data allow the user to find flaws in the soil, leaving to the geologist the decision whether the well

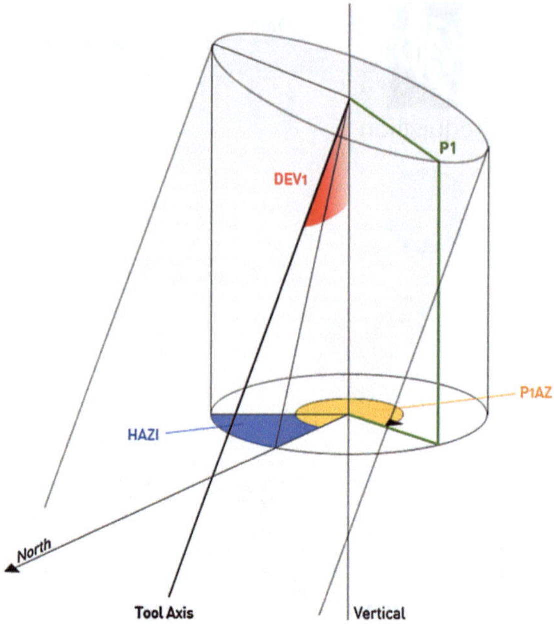

Fig. 13.4 A 3D sketch of the Caliper tool orientation

is feasible or not, and possible actions to minimize its usage risk. Figure 13.5 shows an example of a caliper sample file, as it is provided by the tool.

Table 13.1 shows an example of a single line data obtained from a file of the Caliper Tool.

13.3.2 Preprocessing

The borehole data is stored in a file and then read and loaded into the main memory. The reading process takes a while, due the large file size and the transformations that are applied to transform it into tridimensional data. For example, a 4 Km borehole is stored in a file of about 180 MB of raw data that must still be read and converted into the three-dimensional borehole geometry.

From a usability point of view, it is not convenient that the system stays locked until all the geometry is ready for rendering. Hence, we propose in our process to show some visual information in the screen as soon as possible. This is done by combining two working threads, one responsible for reading and transforming the data and the other that takes this data and builds the mesh geometry.

This threading part is managed with C++ Boost's thread module [9], a portable library which allows execution of multiple threads with shared data and provides functions for thread management and synchronization.

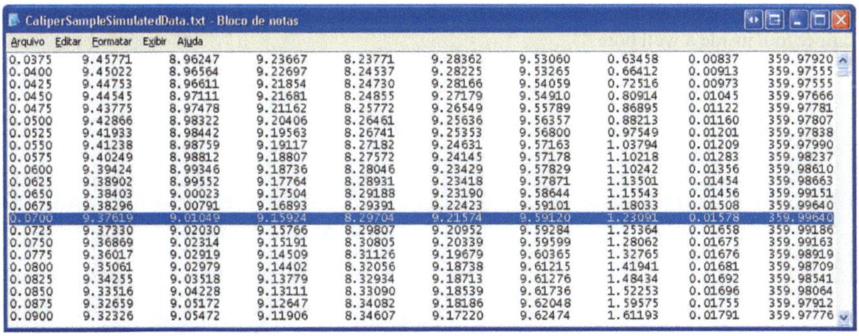

0.0375	9.45771	8.96247	9.23667	8.23771	9.28362	9.53060	0.63458	0.00837	359.97920
0.0400	9.45022	8.96564	9.22697	8.24537	9.28225	9.53265	0.66412	0.00913	359.97555
0.0425	9.44753	8.96611	9.21854	8.24730	9.28166	9.54059	0.72516	0.00973	359.97555
0.0450	9.44545	8.97111	9.21681	8.24855	9.27179	9.54910	0.80914	0.01045	359.97666
0.0475	9.43775	8.97478	9.21162	8.25772	9.26549	9.55789	0.86895	0.01122	359.97781
0.0500	9.42866	8.98322	9.20406	8.26461	9.25636	9.56357	0.88213	0.01160	359.97807
0.0525	9.41933	8.98442	9.19563	8.26741	9.25353	9.56800	0.97549	0.01201	359.97838
0.0550	9.41238	8.98759	9.19117	8.27182	9.24631	9.57163	1.03794	0.01209	359.97990
0.0575	9.40249	8.98812	9.18807	8.27572	9.24145	9.57178	1.10218	0.01283	359.98237
0.0600	9.39424	8.99346	9.18736	8.28046	9.23429	9.57829	1.10242	0.01356	359.98610
0.0625	9.38902	8.99552	9.17764	8.28931	9.23418	9.57871	1.13501	0.01454	359.98663
0.0650	9.38403	9.00023	9.17504	8.29188	9.23190	9.58644	1.15543	0.01456	359.99151
0.0675	9.38296	9.00791	9.16893	8.29391	9.22423	9.59101	1.18033	0.01508	359.99640
0.0700	9.37619	9.01049	9.15924	8.29704	9.21574	9.59120	1.23091	0.01578	359.99640
0.0725	9.37330	9.02030	9.15766	8.29807	9.20952	9.59284	1.25364	0.01658	359.99186
0.0750	9.36869	9.02314	9.15191	8.30805	9.20339	9.59599	1.28062	0.01675	359.99163
0.0775	9.36017	9.02919	9.14509	8.31126	9.19679	9.60365	1.32765	0.01676	359.98919
0.0800	9.35061	9.02979	9.14402	8.32056	9.18738	9.61215	1.41941	0.01681	359.98709
0.0825	9.34255	9.03518	9.13779	8.32934	9.18713	9.61276	1.48434	0.01692	359.98541
0.0850	9.33516	9.04228	9.13111	8.33090	9.18539	9.61736	1.52253	0.01696	359.98064
0.0875	9.32659	9.05172	9.12647	8.34082	9.18186	9.62048	1.59575	0.01755	359.97912
0.0900	9.32326	9.05472	9.11906	8.34607	9.17220	9.62474	1.61193	0.01791	359.97776

Fig. 13.5 Sample data obtained from Caliper tool

Table 13.1 A sample of some caliper fundamental measures data

Data	Sample value	Definition
Depth measure	0.07 m	Measure's depth
Arm 01	9.37619 in	Length of Arm 01
Arm 02	9.01049 in	Length of Arm 02
Arm 03	9.15924 in	Length of Arm 03
Arm 04	8.29704 in	Length of Arm 04
Arm 05	9.21574 in	Length of Arm 05
Arm 06	9.59120 in	Length of Arm 06
DEVI	1.23091°	Deviation from the surface
HAZI	0.01578°	Projection of the borehole axis in the horizontal plane
P1AZ	359.99640°	Projection of pad 1 in the horizontal plane

13.3.3 Math

At this stage we implement one of the most important parts of the visualization tool, which is the transformation process of the 2D data obtained by the Caliper tool into 3D data. This effort is divided in 6 steps as can be see on the following subtopics.

13.3.3.1 Transformation of the Caliper Arms Length into a 2D Data

The Caliper data file stores the arms' length as a single value that represents the distance from the tool center and the arm pad. This data must be transformed into 2D coordinates (and later 3D) in order to build the borehole geometry.

For the sake of simplicity, the tool axes are located in the center of the Cartesian coordinate system and the direction of the first arm coincides with the positive X-axis, as shown in Fig. 13.6. The arms can be described in polar coordinates, so each coordinate for the other arms can be obtained from equations in Table 13.2.

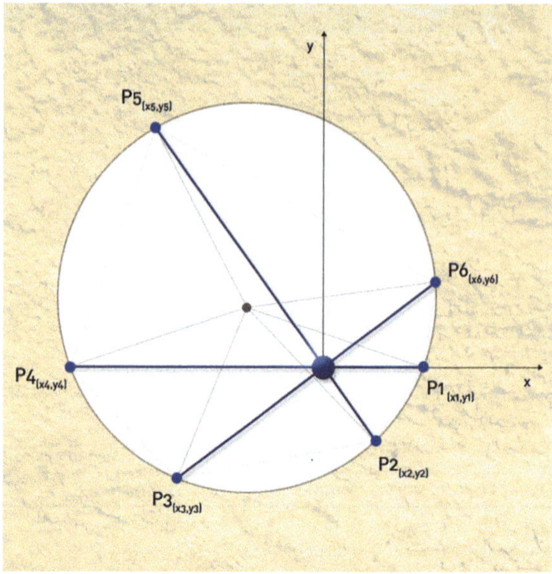

Fig. 13.6 Top View of the Caliper tool inside a well bore

Table 13.2 How to calculate the X and Y coordinates

Data	X-Value formula	Y-Value formula
Arm C1	P1 cos 0°	P1 sin 0°
Arm C2	P2 cos 300°	P2 sin 300°
Arm C3	P3 cos 240°	P3 sin 240°
Arm C4	P4 cos 180°	P4 sin 180°
Arm C5	P5 cos 120°	P5 sin 120°
Arm C6	P6 cos 60°	P6 sin 60°

13.3.3.2 Interpolation of the Data with Splines to Obtain a Section with the Desirable Resolution

Given six 2D control points (obtained from the Caliper's arm data) instead of simple connect the points (red lines in Fig. 13.7) a smooth curve is created (black lines in the same figure) to interpolate them using a closed natural cubic splines algorithm. An interesting demonstration can be found in [10].

13.3.3.3 Inclusion of the Rotation in P1AZ

The interpolated points are rotated around the section center in the Y-axis. The rotation angle is defined by the P1AZ value read by the sonde. This aligns the first arm with the earth's magnetic north.

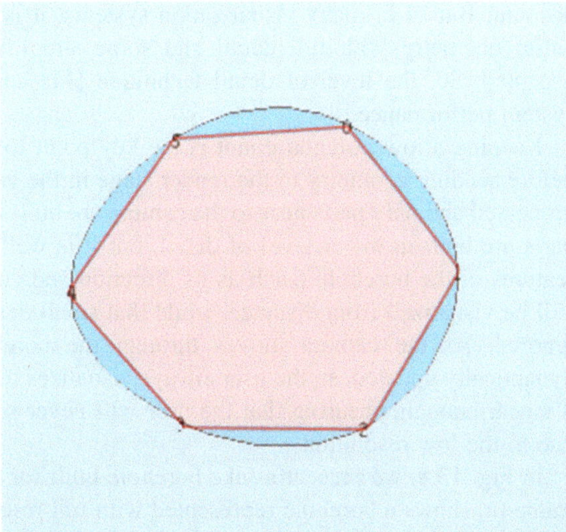

Fig. 13.7 Drawn in *red* is the original data acquired from the six-arms caliper control points. In *blue* is an approximation representation created with an interpolation of a closed natural cubic splines

13.3.3.4 Inclusion of the Rotation in DEVI

DEVI value defines the sonde deviation from the surface. Combined with HAZI, it's used to determine the tool's inclination and direction.

13.3.3.5 Inclusion of the Rotation in HAZI

HAZI tells the borehole angle with the earth's magnetic north. This value, combined with DEVI, allows finding out tool's inclination and direction.

13.4 Optimization

Optimizations are necessary in order to keep the visualization system running in real-time at acceptable frame rates. This section describes some techniques used for this purpose for this work.

13.4.1 Level of Detail

Due to the common large size of the boreholes, it is not practical to render its entire geometry in every frame. This could easily lead to GPU memory overflow and even if memory is not an issue, processing time should still be taken into account. But as in many visualization systems, it is not mandatory to render the entire geometry with full detail and some simplification for distant objects is accepted. So, the level of detail technique [11] can be applied to increase the system performance.

Keeping a low polygon count is the key point to achieve high frame rates, so before sending geometry to the render stage in the graphics pipeline, the model is processed and only parts near to the camera are built in high level of detail. Distant parts are built in lower level of detail, but it is well-balanced enough that major features of the borehole (such as its direction and curvature) are not lost and can still be visualized from distance, while that small details from distant sections are ignored. As the camera moves through the borehole, the level of detail is dynamically updated, so the user always visualizes the high-resolution mesh when it's near enough, ensuring that the user will never see problems on the geometry due to the low resolution.

In Fig. 13.8, we present a fake borehole built for demonstration only. The first frame (a) shows a borehole represented with full resolution. Here, all the borehole data is used for rendering. The next frame (b) shows how the borehole should be represented with a lower resolution. The sections are not interpolated, so each one is in fact a hexagon, and only a small fraction of the original data is used to build the geometry, this maintains the overall shape of the borehole, but small variations are lost. The last frame (c) mixes these two visualizations and shows a high-resolution borehole for parts that are near to the camera. As the camera moves away, lower detail level is required, while the viewer still has the impression that the geometry is unchanged. This happens because, different from this example, in the actual system we only apply the level of detail for parts that are really far away from the camera and smaller geometry changes may go unnoticed.

13.4.2 Concurrent Mesh Manipulation and Parallelism

As described in Sect. 13.3.2—Preprocessing, the borehole data is loaded without locking the visualization window. Whenever a thread finishes loading a part of the borehole, it is immediately sent to the rendering without major performance costs. The process for changing the level of detail after the model is fully loaded uses a similar technique.

For memory space reasons, only one mesh for a given part of the borehole is stored in the GPU memory at a time. Whenever the level of details changes, a new mesh with a different level of detail is again built by the CPU and sent to the GPU. The older mesh is still rendered until the new mesh is ready, so a gap is never formed while the meshes are changed. The building process of a new mesh is done by a concurrent thread, so no slow down is noticed. When it's ready for rendering, the older mesh is removed from memory and the new one becomes visible.

The borehole is divided in a variable number of meshes depending on the borehole size. Anytime the level of detail should be changed, i.e., when the camera

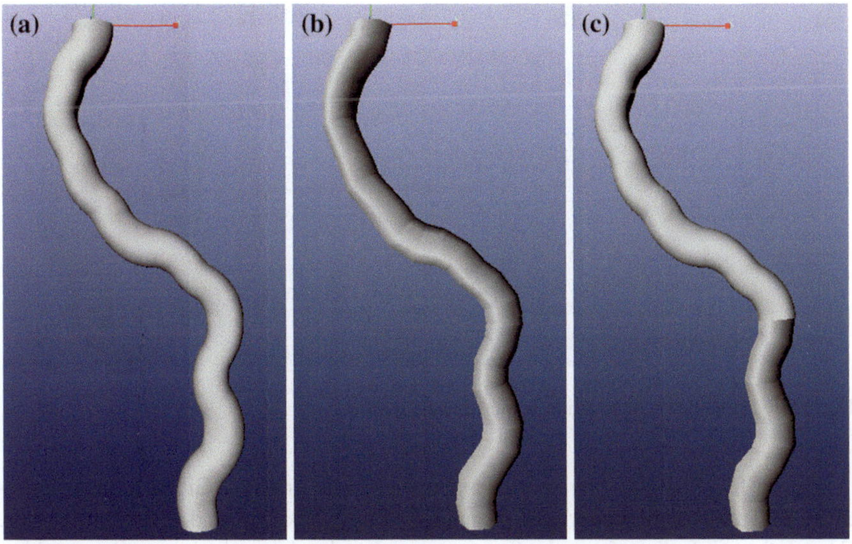

Fig. 13.8 A fake borehole three-dimensional mesh with different levels of detail

moves, only the closer meshes are updated, not the borehole model. This is a different approach from the commonly used technique, where the model is fully changed for each level of detail.

13.5 Rendering

Once the data is fully loaded into the main memory and processed, it's possible to reconstruct a three-dimensional representation of the borehole. As previously described, some transformations are applied to caliper arm's points in order to transform them from a bi-dimensional plane to a three-dimensional space. The next step is to combine each section and create a 3D model of the borehole that can be used for visualization and analysis. If this combination had not been done, the model would simply look like as a simple stack of slices instead of a reliable three-dimensional representation of the borehole.

The representation of the borehole as a series of connected sections makes the usage of triangle strips [12] a natural choice for building the geometry. The connection between two adjacent sections is built with a single triangle strip, instead of a collection of independent triangles. This results in lower memory usage by the GPU and better performance. Figure 13.9 shows how we start from points obtained from the spline interpolation (a) and build the borehole geometry (b and c).

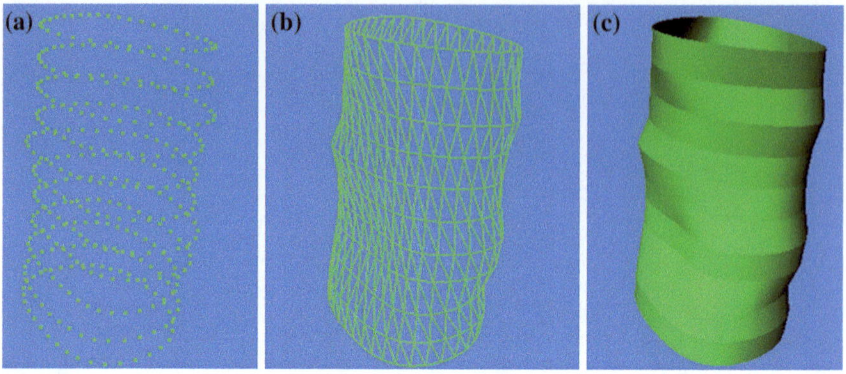

Fig. 13.9 Borehole reconstruction using triangle strips

Figure 13.10 shows one simple example of this process. Given a pair of sections, let's call A the upper section indicated by black dots and B the white ones. From left to right (i.e., from the first arm or control point to the last one), we add to the triangle strip the first vertex from A and B, in this order (points 1 and 7 in the figure). Then, we move to the next pair and add the new points from A and B (2 and 8), and so on until all vertex from both sections are added. The first two points are added again in the end, so the strip closes the section. Each new point added after the first pair results in the creation of a new triangle (1-7-2 and 7-2-8, for example). When the process is over, a small mesh connecting two sections is created. Applying these steps to all sections it results in a 3D representation of the borehole. Listing 13.1 describes a pseudo-code for this process.

13.6 Analysis and Information

Study of borehole stability requires not only a deep knowledge of problems' sources, but it is accurate identification and modeling. When a borehole is drilled in a nonisotropic prestressed rock, we can note a failure, which is a result from the reorientation of the stress field and the stress concentration around the borehole wall. The stability of an oil reservoir is a challenge for specialists of the petroleum industry; therefore, a correct analysis of this question can reduce the perforation cost significantly. It can be considering that a considerable time expense to perforate a hole is related to the analysis of its stability, which means an annually high cost.

The caliper tool is used in this purpose for gathering data such as resistivity and acoustic measures of the borehole wall, as well as diameter for each section. These data can be used to identify problematic points where tension could lead to the borehole's collapse.

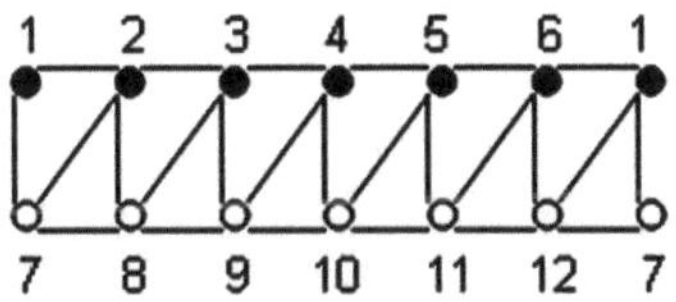

Fig. 13.10 Sample triangle strip connecting two sections

```
for i = 0; i < sections.size - 1; i++
    A = sections[i]
    B = sections[i+1]
    for j = 0; j < A.vertexCount; j++
        addVertex (A[j])
        addVertex (B[j])
```

Listing 13.1 Pseudocode for creating the triangle strips

The three-dimensional model of the borehole allows the user to visualize the possible problematic points and have access to collected data for the sections. For each section, the user is able to request many calculations, such as sections diameters relation, section angle, and cut type classification. This combination of data and visual information allows an easier way to identify whether the borehole is stable and which actions should be taken.

13.7 The Implementation of the Three-Dimensional Visualization System

The visualization system, named "GeoPoço 3D", was written in C++ with an OpenGL-based graphics framework. An initial prototype was made using Irrlicht [13], an open-source 3D engine that provides a certain level of abstraction of hardware resources by its API and great performance. A new version is under development using Open Inventor [14] instead of Irrlicht.

The tests were carried out on a desktop with the Windows XP Professional SP2 operating system was installed, with a Core 2 Duo processor E7500 2.93 Ghz and 2 GB of RAM. The graphic card used it was an NVidia GeForce 9600 GT with 512 MB of RAM and 64 stream processors. The former prototype.

13.8 Results

A key point in this visualization system is that very large boreholes can be analyzed without visible performance loss. Should the borehole be rendered entirely in high-resolution, the GPU would quickly run out of video memory and its processing power is very likely to not be able to deal with this data flow.

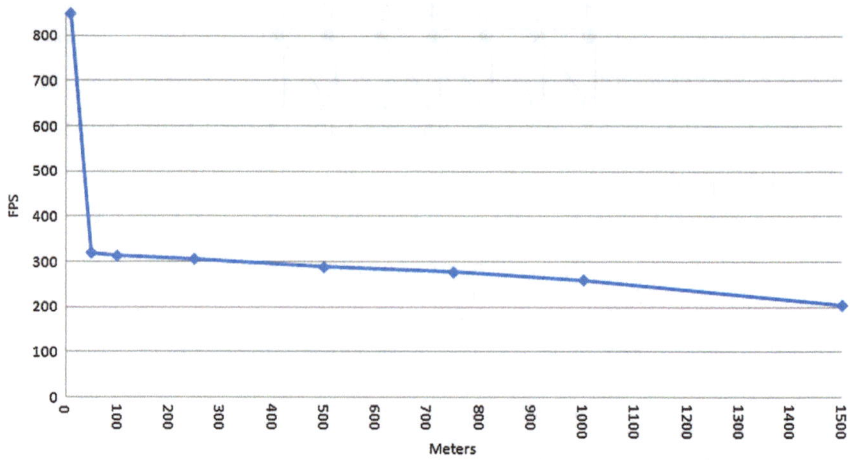

Fig. 13.11 Comparison of performance of the FPS rate versus the growth of the well length

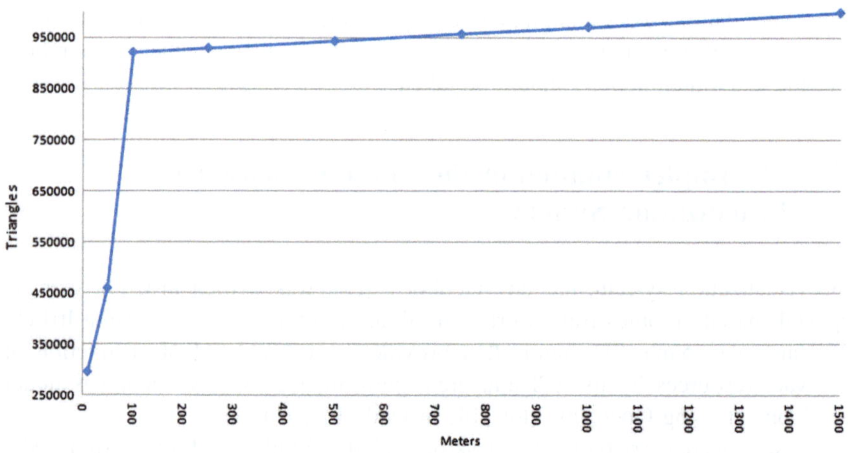

Fig. 13.12 Comparison of the growth of the well length versus the amount of triangles

The level of detail technique allows the borehole to be fully visualized, but only the nearest parts are rendered with full detail level. Distant parts are built with very simple geometry that still looks like the original borehole, but lacks detail.

This technique allows the visualization of very large boreholes without major performance issues. As the borehole gets bigger, only a small fraction of its structure is added to the scene graph, so the frame rate tends to be stable. Figure 13.11 demonstrates how the system behaves as the borehole size increases. A 10 meter borehole runs at more than 800 frames per second. In this case, no level of detail was applied. The frame rate drops more than 2.5 times for a borehole 10 times larger (100 m), with level of detail already applied. One should expect that when we double the borehole size, the frame rate should drop 5 times and so on. But with the level of detail applied, the triangle count does not grow much fast and the frame rate becomes stable.

The system is scalable for visualization of very large boreholes. Since it's estimated that specialists will analyze sections of 500 m or less at a time, the current performance is more than enough for a completely smooth analysis.

Figure 13.12 shows how the number of triangles does not increase as fast as the borehole length, consequence of the level of detail used to render the scene, what means that only the necessary triangles are sent to GPU memory. This allows larger sections and higher resolutions without significant rate loss.

13.9 Conclusions

In this chapter, the authors presented a framework for real-time visualization and geometry reconstruction of large oil and gas boreholes.

The visualization of a borehole is a relevant issue for the oil and gas industry as well for the productivity of the field. If you know where the problem can occur or where it is better to drill you can increase your production.

The tool proved to be able to scale up for really huge wells since in all the instances clearly show that even on the wellbores with thousands of meters the system provides an interactive rate of at least 200 frames per second for all the well-visualization tasks like moving, zooming, rotation, and translation.

The tool can help you spend less time analyzing wellbore stability issues, helping decisions, reducing project risks, drilling costs, and increasing the project profit.

However, there are still many features to be implemented as you can see on the future work section.

13.10 Future Works

At present, we are only using a simple interface with a mouse and a 3D view. The next aim is to provide a better visualization of the well using a touch surface table, controllers, and 3D glasses. As future work, we also intend to target it at further development, increasing interaction and automated diagnosis of failures on the well.

Acknowledgments The authors would like to acknowledge Petrobras Oil for the financial support.

References

1. Jarosinski, M., Zoback, M.D.: Comparison of six-arm caliper and borehole televiewer data for detection of stress induced wellbore breakouts: application to six wheels. Polish Carpathians, pp. F8-1–F8-23+12 figures (1998)
2. Reinsch, C.: Smoothing by spline functions. Numer. Math. **10**, 177–183 (1999). In: Proceedings of the XII SIBGRAPI pp. 101–104 (1967)
3. Aadnoy, B.S., Balkema, A.A.: Modern well design. Rotterdam. ISBN 90 54106336 (1996)
4. Jiménez, J.C, Lara, L.V. et al.: Geomechanical wellbore stability modelling of exploratory wells—study case at middle magdalena basin. C.T.F Cienc. Technol. Futuro **3**(3), 85–102 (2007)
5. Barton, C., Zoback, M.D.: Stress perturbations associated with active faults penetrated by boreholes: possible evidence for near-complete stress drop and a new technique for stress magnitude measurements. J. Geophys. Res. **99**, 9,373–9,390 (1994)
6. Peska, P., Zoback, M.D.: Compressive and tensile failure of inclined well-bores and direct determination of in situ stress and rock strength. J. Geophys. Res. **100**, 12,791–12,811 (1995)
7. Leta, F.R., Souza, M., Clua, E., Biondi, M., Pacheco, T.: Computational system to help the stress analysis around boreholes in petroleum industry. In: Proceedings of the ECCOMAS 2008, Venice (2008)
8. Dembogurski, B., Clua, E., Leta, F.R., Procedural terrain generation at GPU level with marching cubes. In: VII Brazilian Symposium of Games and Digital Entertainment—Computing Track. Proceedings of the VII Brazilian Symposium of Games and Digital Entertainment—Short Papers. Porto Alegre: Sociedade Brasileira da Computação pp. 37–40 (2008)
9. Williams, A.: Boost C++ Libraries Chapter 22—Thread http://www.boost.org/doc/libs/1_43_0/doc/html/thread.html (2008)
10. Lambert, T.: Closed natural cubic splines. www.cse.unsw.edu.au/lambert/splines/natcubicclosed.html
11. Luebke, D., Reddy, M., et al.: Level of detail for 3D graphics. Morgan Kaufmann, San Francisco 432 p (2003)
12. Schroeder, W.J.: Modeling of surfaces employing polygon strips. http://patft1.uspto.gov/netacgi/nphParser?Sect1=PTO1&Sect2=HITOFF&d=PALL&p=1&u=%252Fnetahtml%252FPTO%252Fsrchnum.htm&r=1&f=G&l=50&s1=5,561,749.PN.&OS=PN/5,561,749 (1996)
13. Gebhardt, N.: Irrlicht Engine. http://irrlicht.sourceforge.net/ (2010)
14. Open Inventor http://www.vsg3d.com/vsg_prod_openinventor.php (2010)

MIX
Papier aus verantwortungsvollen Quellen
Paper from responsible sources
FSC® C105338

If you have any concerns about our products,
you can contact us on
ProductSafety@springernature.com

In case Publisher is established outside the EU,
the EU authorized representative is:
Springer Nature Customer Service Center GmbH
Europaplatz 3, 69115 Heidelberg, Germany

Printed by Libri Plureos GmbH
in Hamburg, Germany